FREE Test Taking Tips DVD Offer

To help us better serve you, we have developed a Test Taking Tips DVD that we would like to give you for FREE. **This DVD covers world-class test taking tips that you can use to be even more successful when you are taking your test.**

All that we ask is that you email us your feedback about your study guide. Please let us know what you thought about it – whether that is good, bad or indifferent.

To get your **FREE Test Taking Tips DVD**, email freedvd@studyguideteam.com with "FREE DVD" in the subject line and the following information in the body of the email:

> a. The title of your study guide.
>
> b. Your product rating on a scale of 1-5, with 5 being the highest rating.
>
> c. Your feedback about the study guide. What did you think of it?
>
> d. Your full name and shipping address to send your free DVD.

If you have any questions or concerns, please don't hesitate to contact us at freedvd@studyguideteam.com.

Thanks again!

ACCUPLACER
Math Study Guide

ACCUPLACER Math Test Prep and
Practice Exam Questions [2nd Edition]

TPB Publishing

Interested in buying more than 10 copies of our product? Contact us about bulk discounts:
bulkorders@studyguideteam.com

ISBN 13: 9781628456929
ISBN 10: 1628456922

Test Prep Books!!!

Table of Contents

Quick Overview

As you draw closer to taking your exam, effective preparation becomes more and more important. Thankfully, you have this study guide to help you get ready. Use this guide to help keep your studying on track and refer to it often.

This study guide contains several key sections that will help you be successful on your exam. The guide contains tips for what you should do the night before and the day of the test. Also included are test-taking tips. Knowing the right information is not always enough. Many well-prepared test takers struggle with exams. These tips will help equip you to accurately read, assess, and answer test questions.

A large part of the guide is devoted to showing you what content to expect on the exam and to helping you better understand that content. In this guide are practice test questions so that you can see how well you have grasped the content. Then, answer explanations are provided so that you can understand why you missed certain questions.

Don't try to cram the night before you take your exam. This is not a wise strategy for a few reasons. First, your retention of the information will be low. Your time would be better used by reviewing information you already know rather than trying to learn a lot of new information. Second, you will likely become stressed as you try to gain a large amount of knowledge in a short amount of time. Third, you will be depriving yourself of sleep. So be sure to go to bed at a reasonable time the night before. Being well-rested helps you focus and remain calm.

Be sure to eat a substantial breakfast the morning of the exam. If you are taking the exam in the afternoon, be sure to have a good lunch as well. Being hungry is distracting and can make it difficult to focus. You have hopefully spent lots of time preparing for the exam. Don't let an empty stomach get in the way of success!

When travelling to the testing center, leave earlier than needed. That way, you have a buffer in case you experience any delays. This will help you remain calm and will keep you from missing your appointment time at the testing center.

Be sure to pace yourself during the exam. Don't try to rush through the exam. There is no need to risk performing poorly on the exam just so you can leave the testing center early. Allow yourself to use all of the allotted time if needed.

Remain positive while taking the exam even if you feel like you are performing poorly. Thinking about the content you should have mastered will not help you perform better on the exam.

Once the exam is complete, take some time to relax. Even if you feel that you need to take the exam again, you will be well served by some down time before you begin studying again. It's often easier to convince yourself to study if you know that it will come with a reward!

Test-Taking Strategies

1. Predicting the Answer

When you feel confident in your preparation for a multiple-choice test, try predicting the answer before reading the answer choices. This is especially useful on questions that test objective factual knowledge. By predicting the answer before reading the available choices, you eliminate the possibility that you will be distracted or led astray by an incorrect answer choice. You will feel more confident in your selection if you read the question, predict the answer, and then find your prediction among the answer choices. After using this strategy, be sure to still read all of the answer choices carefully and completely. If you feel unprepared, you should not attempt to predict the answers. This would be a waste of time and an opportunity for your mind to wander in the wrong direction.

2. Reading the Whole Question

Too often, test takers scan a multiple-choice question, recognize a few familiar words, and immediately jump to the answer choices. Test authors are aware of this common impatience, and they will sometimes prey upon it. For instance, a test author might subtly turn the question into a negative, or he or she might redirect the focus of the question right at the end. The only way to avoid falling into these traps is to read the entirety of the question carefully before reading the answer choices.

3. Looking for Wrong Answers

Long and complicated multiple-choice questions can be intimidating. One way to simplify a difficult multiple-choice question is to eliminate all of the answer choices that are clearly wrong. In most sets of answers, there will be at least one selection that can be dismissed right away. If the test is administered on paper, the test taker could draw a line through it to indicate that it may be ignored; otherwise, the test taker will have to perform this operation mentally or on scratch paper. In either case, once the obviously incorrect answers have been eliminated, the remaining choices may be considered. Sometimes identifying the clearly wrong answers will give the test taker some information about the correct answer. For instance, if one of the remaining answer choices is a direct opposite of one of the eliminated answer choices, it may well be the correct answer. The opposite of obviously wrong is obviously right! Of course, this is not always the case. Some answers are obviously incorrect simply because they are irrelevant to the question being asked. Still, identifying and eliminating some incorrect answer choices is a good way to simplify a multiple-choice question.

4. Don't Overanalyze

Anxious test takers often overanalyze questions. When you are nervous, your brain will often run wild, causing you to make associations and discover clues that don't actually exist. If you feel that this may be a problem for you, do whatever you can to slow down during the test. Try taking a deep breath or counting to ten. As you read and consider the question, restrict yourself to the particular words used by the author. Avoid thought tangents about what the author *really* meant, or what he or she was *trying* to say. The only things that matter on a multiple-choice test are the words that are actually in the question. You must avoid reading too much into a multiple-choice question, or supposing that the writer meant something other than what he or she wrote.

5. No Need for Panic

It is wise to learn as many strategies as possible before taking a multiple-choice test, but it is likely that you will come across a few questions for which you simply don't know the answer. In this situation, avoid panicking. Because most multiple-choice tests include dozens of questions, the relative value of a single wrong answer is small. As much as possible, you should compartmentalize each question on a multiple-choice test. In other words, you should not allow your feelings about one question to affect your success on the others. When you find a question that you either don't understand or don't know how to answer, just take a deep breath and do your best. Read the entire question slowly and carefully. Try rephrasing the question a couple of different ways. Then, read all of the answer choices carefully. After eliminating obviously wrong answers, make a selection and move on to the next question.

6. Confusing Answer Choices

When working on a difficult multiple-choice question, there may be a tendency to focus on the answer choices that are the easiest to understand. Many people, whether consciously or not, gravitate to the answer choices that require the least concentration, knowledge, and memory. This is a mistake. When you come across an answer choice that is confusing, you should give it extra attention. A question might be confusing because you do not know the subject matter to which it refers. If this is the case, don't eliminate the answer before you have affirmatively settled on another. When you come across an answer choice of this type, set it aside as you look at the remaining choices. If you can confidently assert that one of the other choices is correct, you can leave the confusing answer aside. Otherwise, you will need to take a moment to try to better understand the confusing answer choice. Rephrasing is one way to tease out the sense of a confusing answer choice.

7. Your First Instinct

Many people struggle with multiple-choice tests because they overthink the questions. If you have studied sufficiently for the test, you should be prepared to trust your first instinct once you have carefully and completely read the question and all of the answer choices. There is a great deal of research suggesting that the mind can come to the correct conclusion very quickly once it has obtained all of the relevant information. At times, it may seem to you as if your intuition is working faster even than your reasoning mind. This may in fact be true. The knowledge you obtain while studying may be retrieved from your subconscious before you have a chance to work out the associations that support it. Verify your instinct by working out the reasons that it should be trusted.

8. Key Words

Many test takers struggle with multiple-choice questions because they have poor reading comprehension skills. Quickly reading and understanding a multiple-choice question requires a mixture of skill and experience. To help with this, try jotting down a few key words and phrases on a piece of scrap paper. Doing this concentrates the process of reading and forces the mind to weigh the relative importance of the question's parts. In selecting words and phrases to write down, the test taker thinks about the question more deeply and carefully. This is especially true for multiple-choice questions that are preceded by a long prompt.

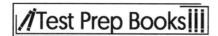

9. Subtle Negatives

One of the oldest tricks in the multiple-choice test writer's book is to subtly reverse the meaning of a question with a word like *not* or *except*. If you are not paying attention to each word in the question, you can easily be led astray by this trick. For instance, a common question format is, "Which of the following is…?" Obviously, if the question instead is, "Which of the following is not…?," then the answer will be quite different. Even worse, the test makers are aware of the potential for this mistake and will include one answer choice that would be correct if the question were not negated or reversed. A test taker who misses the reversal will find what he or she believes to be a correct answer and will be so confident that he or she will fail to reread the question and discover the original error. The only way to avoid this is to practice a wide variety of multiple-choice questions and to pay close attention to each and every word.

10. Reading Every Answer Choice

It may seem obvious, but you should always read every one of the answer choices! Too many test takers fall into the habit of scanning the question and assuming that they understand the question because they recognize a few key words. From there, they pick the first answer choice that answers the question they believe they have read. Test takers who read all of the answer choices might discover that one of the latter answer choices is actually *more* correct. Moreover, reading all of the answer choices can remind you of facts related to the question that can help you arrive at the correct answer. Sometimes, a misstatement or incorrect detail in one of the latter answer choices will trigger your memory of the subject and will enable you to find the right answer. Failing to read all of the answer choices is like not reading all of the items on a restaurant menu: you might miss out on the perfect choice.

11. Spot the Hedges

One of the keys to success on multiple-choice tests is paying close attention to every word. This is never truer than with words like almost, most, some, and sometimes. These words are called "hedges" because they indicate that a statement is not totally true or not true in every place and time. An absolute statement will contain no hedges, but in many subjects, the answers are not always straightforward or absolute. There are always exceptions to the rules in these subjects. For this reason, you should favor those multiple-choice questions that contain hedging language. The presence of qualifying words indicates that the author is taking special care with his or her words, which is certainly important when composing the right answer. After all, there are many ways to be wrong, but there is only one way to be right! For this reason, it is wise to avoid answers that are absolute when taking a multiple-choice test. An absolute answer is one that says things are either all one way or all another. They often include words like *every*, *always*, *best*, and *never*. If you are taking a multiple-choice test in a subject that doesn't lend itself to absolute answers, be on your guard if you see any of these words.

12. Long Answers

In many subject areas, the answers are not simple. As already mentioned, the right answer often requires hedges. Another common feature of the answers to a complex or subjective question are qualifying clauses, which are groups of words that subtly modify the meaning of the sentence. If the question or answer choice describes a rule to which there are exceptions or the subject matter is complicated, ambiguous, or confusing, the correct answer will require many words in order to be expressed clearly and accurately. In essence, you should not be deterred by answer choices that seem excessively long. Oftentimes, the author of the text will not be able to write the correct answer without

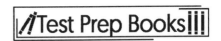

offering some qualifications and modifications. Your job is to read the answer choices thoroughly and completely and to select the one that most accurately and precisely answers the question.

13. Restating to Understand

Sometimes, a question on a multiple-choice test is difficult not because of what it asks but because of how it is written. If this is the case, restate the question or answer choice in different words. This process serves a couple of important purposes. First, it forces you to concentrate on the core of the question. In order to rephrase the question accurately, you have to understand it well. Rephrasing the question will concentrate your mind on the key words and ideas. Second, it will present the information to your mind in a fresh way. This process may trigger your memory and render some useful scrap of information picked up while studying.

14. True Statements

Sometimes an answer choice will be true in itself, but it does not answer the question. This is one of the main reasons why it is essential to read the question carefully and completely before proceeding to the answer choices. Too often, test takers skip ahead to the answer choices and look for true statements. Having found one of these, they are content to select it without reference to the question above. Obviously, this provides an easy way for test makers to play tricks. The savvy test taker will always read the entire question before turning to the answer choices. Then, having settled on a correct answer choice, he or she will refer to the original question and ensure that the selected answer is relevant. The mistake of choosing a correct-but-irrelevant answer choice is especially common on questions related to specific pieces of objective knowledge. A prepared test taker will have a wealth of factual knowledge at his or her disposal, and should not be careless in its application.

15. No Patterns

One of the more dangerous ideas that circulates about multiple-choice tests is that the correct answers tend to fall into patterns. These erroneous ideas range from a belief that B and C are the most common right answers, to the idea that an unprepared test-taker should answer "A-B-A-C-A-D-A-B-A." It cannot be emphasized enough that pattern-seeking of this type is exactly the WRONG way to approach a multiple-choice test. To begin with, it is highly unlikely that the test maker will plot the correct answers according to some predetermined pattern. The questions are scrambled and delivered in a random order. Furthermore, even if the test maker was following a pattern in the assignation of correct answers, there is no reason why the test taker would know which pattern he or she was using. Any attempt to discern a pattern in the answer choices is a waste of time and a distraction from the real work of taking the test. A test taker would be much better served by extra preparation before the test than by reliance on a pattern in the answers.

FREE DVD OFFER

Don't forget that doing well on your exam includes both understanding the test content and understanding how to use what you know to do well on the test. We offer a completely FREE Test Taking Tips DVD that covers world class test taking tips that you can use to be even more successful when you are taking your test.

All that we ask is that you email us your feedback about your study guide. To get your **FREE Test Taking Tips DVD**, email freedvd@studyguideteam.com with "FREE DVD" in the subject line and the following information in the body of the email:

- The title of your study guide.
- Your product rating on a scale of 1-5, with 5 being the highest rating.
- Your feedback about the study guide. What did you think of it?
- Your full name and shipping address to send your free DVD.

Introduction to the ACCUPLACER Exam

Function of the Test

ACCUPLACER is an adaptive, computerized test offered by the College Board and used by some colleges and high schools to determine placement of students in programs appropriate to the students' skill level. The test is offered nation-wide, at any college or high school that chooses to use it. Test-takers are almost always students of schools that use the ACCUPLACER in their course selection and placement efforts. Schools can also use the ACCUPLACER to assess students' skill levels and identify specific areas in which the students need improvement. Scores are generally used only by the college or high school the student is already attending for placement and instruction at that school.

According to the College Board, more than 1,500 secondary and post-secondary institutions are currently using the ACCUPLACER. Over 7.5 million ACCUPLACER tests are administered in a typical year. The College Board recommends that ACCUPLACER scores be used in conjunction with other variables including high school GPA, the number of years a student has taken coursework in a particular subject area, other test scores such as the SAT or ACT, as well as non-cognitive information such as motivation, family support, and time management skills, in order to place students in courses of appropriate difficulty.

Test Administration

ACCUPLACER is offered by computer, usually by the school that wants to use its results, whenever a student makes an appointment with the school to take it. In cases where a student is not able to take the test at the student's school, arrangements can sometimes be made to take the test at a more convenient location on another school's campus.

Fees for taking the ACCUPLACER vary from school to school. Students sometimes do not pay a fee at all; rather, the school pays the College Board for the right to administer the test and then students take it for free. Students may retake the ACCUPLACER at the discretion of the school administering the test and using its results. Students with documented disabilities can make arrangements, through the test center offering the ACCUPLACER, to take the test with appropriate accommodations, including the potential availability of a written version of the test.

Test Format

The core of the ACCUPLACER is five multiple-choice subject area tests: Arithmetic; Quantitative Reasoning, Algebra, and Statistics; Advanced Algebra and Functions, Reading, and Writing. The Writing test has twenty-five questions. It contains one literary passages with five associated questions, and four informational passages that each have five associated questions. A college may ask a student to test in any or all of the five subject areas, depending on the student's and the school's needs. There is also a sixth section, called the WritePlacer, in which students must write a brief essay. Finally, there are four multiple-choice ESL sections of the ACCUPLACER, which schools may ask students, for whom English is a second language, to take.

The multiple-choice sections of the ACCUPLACER are administered by computer. There is no time-limit, so test takers should feel free to work at their own pace. A typical test section might take around 30 minutes to complete. The test is adaptive, meaning it adjusts the difficulty of each question based on

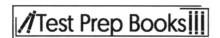

the student's success on the previous questions. The more questions a test-taker gets right, the harder succeeding questions will be and vice-versa. This allows the test to more readily determine test-takers' ability levels without wasting time on questions far above or far below what the test-takers can answer.

Section	Questions	Description
Arithmetic	20	Basic arithmetic and problem solving
Quantitative Reasoning, Algebra, and Statistics	20	Basic algebra, geometry, probability, statistics
Advanced Algebra and Functions	20	Advanced algebra, functions, trigonometry
Reading	20	Information and ideas, rhetoric, synthesis, vocabulary
Writing	25	Expression of ideas, standard English conventions
WritePlacer	1 essay	Effective written communication
ESL- Language Use	20	English grammar
ESL- Listening	20	Understanding spoken English communication
ESL- Reading Skills	20	Comprehension of short written English passages
ESL- Sentence Meaning	20	Understanding the meaning of English sentences

Scoring

The ACCUPLACER subject area tests are scored on a scale from 20 to 120. Schools are free to use the scores for placement as they see fit, given that the difficulty of coursework varies from school to school. A typical community college might separate scores into tiered groups from 50 to 75, from 75 to 99, and from 100 to 120. Scores are typically generated instantly by the computer, but may not be available from the school administering the test until the school has had a chance to review the scores and use them for placement purposes.

Recent Developments

In September 2016, the College Board made "next-generation" ACCUPLACER tests available to schools. Schools have the option of administering the next-generation tests or the older tests, but not both. The new next-generation tests include redesigned reading, writing, and math content, intended to more effectively help schools place students in classes that match their skill level.

Arithmetic

Whole Number Operations

Addition

Addition is the combination of two numbers so their quantities are added together cumulatively. The sign for an addition operation is the + symbol. For example, 9 + 6 = 15. The 9 and 6 combine to achieve a cumulative value, called a **sum**.

Addition holds the commutative property, which means that the order of the numbers in an addition equation can be switched without altering the result. The formula for the commutative property is a + b = b + a. The following examples can demonstrate how the commutative property works:

$$7 = 3 + 4 = 4 + 3 = 7$$

$$20 = 12 + 8 = 8 + 12 = 20$$

Addition also holds the **associative property**, which means that the grouping of numbers does not matter in an addition problem. In other words, the presence or absence of parentheses is irrelevant. The formula for the associative property is (a + b) + c = a + (b + c). Here are some examples of the associative property at work:

$$30 = (6 + 14) + 10 = 6 + (14 + 10) = 30$$

$$35 = 8 + (2 + 25) = (8 + 2) + 25 = 35$$

Subtraction

Subtraction is taking away one number from another, so their quantities are reduced. The sign designating a subtraction operation is the − symbol, and the result is called the **difference.** For example, 9 - 6 = 3. The number *6* detracts from the number *9* to reach the difference *3*.

Unlike addition, subtraction follows neither the commutative nor associative properties. The order and grouping in subtraction impact the result.

$$15 = 22 - 7 \neq 7 - 22 = -15$$

$$3 = (10 - 5) - 2 \neq 10 - (5 - 2) = 7$$

When working through subtraction problems involving larger numbers, it's necessary to regroup the numbers. The following practice problem uses regrouping:

$$\begin{array}{r} 3\,2\,5 \\ -\ 7\,7 \\ \hline \end{array}$$

Here, it is clear that the ones and tens columns for 77 are greater than the ones and tens columns for 325. To subtract this number, one needs to borrow from the tens and hundreds columns. When borrowing from a column, subtracting 1 from the lender column will add 10 to the borrower column:

$$\begin{array}{ccc} 3\text{-}1 & 10\text{+}2\text{-}1 & 10\text{+}5 \\ - & 7 & 7 \\ \hline \end{array} = \begin{array}{ccc} 2 & 11 & 15 \\ - & 7 & 7 \\ \hline 2 & 4 & 8 \end{array}$$

After ensuring that each digit in the top row is greater than the digit in the corresponding bottom row, subtraction can proceed as normal, and the answer is found to be 248.

Multiplication

Multiplication involves adding together multiple copies of a number. It is indicated by an $\times$ symbol or a number immediately outside of a parenthesis. For example:

$$5(8 - 2)$$

The two numbers being multiplied together are called **factors**, and their result is called a **product**. For example, $9 \times 6 = 54$. This can be shown alternatively by expansion of either the 9 or the 6:

$$9 \times 6 = 9 + 9 + 9 + 9 + 9 + 9 = 54$$

$$9 \times 6 = 6 + 6 + 6 + 6 + 6 + 6 + 6 + 6 + 6 = 54$$

Like addition, multiplication holds the commutative and associative properties:

$$115 = 23 \times 5 = 5 \times 23 = 115$$

$$84 = 3 \times (7 \times 4) = (3 \times 7) \times 4 = 84$$

Multiplication also follows the **distributive property**, which allows the multiplication to be distributed through parentheses. The formula for distribution is $a \times (b + c) = ab + ac$. This is clear after the examples:

$$45 = 5 \times 9 = 5(3 + 6) = (5 \times 3) + (5 \times 6) = 15 + 30 = 45$$

$$20 = 4 \times 5 = 4(10 - 5) = (4 \times 10) - (4 \times 5) = 40 - 20 = 20$$

Multiplication becomes slightly more complicated when multiplying numbers with decimals. The easiest way to answer these problems is to ignore the decimals and multiply as if they were whole numbers. After multiplying the factors, a decimal gets placed in the product. The placement of the decimal is determined by taking the cumulative number of decimal places in the factors.

For example:

$$
\begin{array}{r}
0.7 \\
\times\ 3 \\
\hline
2.1
\end{array}
\qquad
\begin{array}{r}
2.6 \\
\times\ 4.2 \\
\hline
10.92
\end{array}
\qquad
\begin{array}{r}
1.5 \\
\times\ 6.4 \\
\hline
9.60
\end{array}
$$

Starting with the first example, the first step is to ignore the decimal and multiply the numbers as though they were whole numbers, which results in a product of 21. The next step is to count the number of digits that follow a decimal (one, in this case). Finally, the decimal place gets moved that many positions to the left, because the factors have only one decimal place. The second example works the same way, except that there are two total decimal places in the factors, so the product's decimal is moved two places over. In the third example, the decimal should be moved over two digits, but the digit zero is no longer needed, so it is erased, and the final answer is 9.6.

Division

Division and multiplication are inverses of each other in the same way that addition and subtraction are opposites. The signs designating the division operation are the ÷ and / symbols. In division, the second number divides into the first.

The number before the division sign is called the **dividend** or, if expressed as a fraction, the **numerator.** For example, in $a \div b$, a is the dividend, while in $\frac{a}{b}$, a is the numerator.

The number after the division sign is called the **divisor** or, if expressed as a fraction, the **denominator.** For example, in $a \div b$, b is the divisor, while in $\frac{a}{b}$, b is the denominator.

Like subtraction, division doesn't follow the commutative property, as it matters which number comes before the division sign, and division doesn't follow the associative or distributive properties for the same reason. For example:

$$\frac{3}{2} = 9 \div 6 \neq 6 \div 9 = \frac{2}{3}$$

$$2 = 10 \div 5 = (30 \div 3) \div 5 \neq 30 \div (3 \div 5) = 30 \div \frac{3}{5} = 50$$

$$25 = 20 + 5 = (40 \div 2) + (40 \div 8) \neq 40 \div (2 + 8) = 40 \div 10 = 4$$

If a divisor doesn't divide into a dividend an integer number of times, whatever is left over is termed the **remainder**. The remainder can be further divided out into decimal form by using long division; however, this doesn't always give a **quotient** with a finite number of decimal places, so the remainder can also be expressed as a fraction over the original divisor.

Division with decimals is similar to multiplication with decimals in that when dividing a decimal by a whole number, one should ignore the decimal and divide as if it was a whole number.

Upon finding the answer, or quotient, the decimal point is inserted at the decimal place equal to that in the dividend.

$$15.75 \div 3 = 5.25$$

When the divisor is a decimal number, both the divisor and dividend get multiplied by 10. This process is repeated until the divisor is a whole number, then one needs to complete the division operation as described above.

$$17.5 \div 2.5 = 175 \div 25 = 7$$

Order of Operations

When solving equations with multiple operations, special rules apply. These rules are known as the **Order of Operations**. The order is as follows: Parentheses, Exponents, Multiplication and Division from left to right, and Addition and Subtraction from left to right. A popular mnemonic device to help remember the order is Please Excuse My Dear Aunt Sally (PEMDAS).

Evaluate the following two problems to understand the Order of Operations:

1) $4 + (3 \times 2)^2 \div 4$

First, solve the operation within the parentheses: $4 + 6^2 \div 4$.
Second, solve the exponent: $4 + 36 \div 4$.
Third, solve the division operation: $4 + 9$.
Fourth, finish the operation with addition for the answer, 13.

2) $2 \times (6 + 3) \div (2 + 1)^2$

$2 \times 9 \div (3)^2$
$2 \times 9 \div 9$
$18 \div 9$
2

Estimation and Rounding

Estimation is finding a value that is close to a solution but is not the exact answer. For example, if there are values in the thousands to be multiplied, then each value can be estimated to the nearest thousand and the calculation performed. This value provides an approximate solution that can be determined very quickly.

When estimating, it's often convenient to **round** a number, which means to give an approximate figure to make it easier to compare amounts or perform mental math. Round up when the digit is 5 or more. The digit used to determine the rounding, and all subsequent digits, become 0, and the selected place value is increased by 1. Here are some examples:

75 rounded to the nearest ten is 80
380 rounded to the nearest hundred is 400
22.697 rounded to the nearest hundredth is 22.70

Round down when rounding on any digit that is below 5. The rounded digit, and all subsequent digits, becomes 0, and the preceding digit stays the same. Here are some examples:

92 rounded to the nearest ten is 90

839 rounded to the nearest hundred is 800

22.643 rounded to the nearest hundredth is 22.64

The same estimation strategies and techniques used when working with standard math problems can be employed when working with real-life situations. Estimation is frequently used in calculations involving money, such as for determining if one has enough money for a purchase, how much one needs to save weekly to buy a desired product, or how much a restaurant bill will sum to.

Applying Operations to Real-World Contexts

Addition and subtraction are **inverse operations**. Adding a number and then subtracting the same number will cancel each other out, resulting in the original number, and vice versa. For example, $8 + 7 - 7 = 8$ and $137 - 100 + 100 = 137$. Similarly, multiplication and division are inverse operations. Therefore, multiplying by a number and then dividing by the same number results in the original number, and vice versa. For example, $8 \times 2 \div 2 = 8$ and $12 \div 4 \times 4 = 12$. Inverse operations are used to work backwards to solve problems. In the case that 7 and a number add to 18, the inverse operation of subtraction is used to find the unknown value ($18 - 7 = 11$). If a school's entire 4th grade was divided evenly into 3 classes each with 22 students, the inverse operation of multiplication is used to determine the total students in the grade ($22 \times 3 = 66$). Additional scenarios involving inverse operations are included in the tables below.

There are a variety of real-world situations in which one or more of the operators is used to solve a problem. The tables below display the most common scenarios.

Addition & Subtraction

	Unknown Result	Unknown Change	Unknown Start
Adding to	5 students were in class. 4 more students arrived. How many students are in class? $5 + 4 = ?$	8 students were in class. More students arrived late. There are now 18 students in class. How many students arrived late? $8 + ? = 18$ Solved by inverse operations $18 - 8 = ?$	Some students were in class early. 11 more students arrived. There are now 17 students in class. How many students were in class early? $? + 11 = 17$ Solved by inverse operations $17 - 11 = ?$
Taking from	15 students were in class. 5 students left class. How many students are in class now? $15 - 5 = ?$	12 students were in class. Some students left class. There are now 8 students in class. How many students left class? $12 - ? = 8$ Solved by inverse operations $8 + ? = 12 \rightarrow 12 - 8 = ?$	Some students were in class. 3 students left class. Then there were 13 students in class. How many students were in class before? $? - 3 = 13$ Solved by inverse operations $13 + 3 = ?$

	Unknown Total	Unknown Addends (Both)	Unknown Addends (One)
Putting together/ taking apart	The homework assignment is 10 addition problems and 8 subtraction problems. How many problems are in the homework assignment? $10 + 8 = ?$	Bobby has $9. How much can Bobby spend on candy and how much can Bobby spend on toys? $9 = ? + ?$	Bobby has 12 pairs of pants. 5 pairs of pants are shorts, and the rest are long. How many pairs of long pants does he have? $12 = 5 + ?$ Solved by inverse operations $12 - 5 = ?$

	Unknown Difference	Unknown Larger Value	Unknown Smaller Value
Comparing	Bobby has 5 toys. Tommy has 8 toys. How many more toys does Tommy have than Bobby? $5 + ? = 8$ Solved by inverse operations $8 - 5 = ?$ Bobby has \$6. Tommy has \$10. How many fewer dollars does Bobby have than Tommy? $10 - 6 = ?$	Tommy has 2 more toys than Bobby. Bobby has 4 toys. How many toys does Tommy have? $2 + 4 = ?$ Bobby has 3 fewer dollars than Tommy. Bobby has \$8. How many dollars does Tommy have? $? - 3 = 8$ Solved by inverse operations $8 + 3 = ?$	Tommy has 6 more toys than Bobby. Tommy has 10 toys. How many toys does Bobby have? $? + 6 = 10$ Solved by inverse operations $10 - 6 = ?$ Bobby has \$5 less than Tommy. Tommy has \$9. How many dollars does Bobby have? $9 - 5 = ?$

Multiplication and Division

	Unknown Product	Unknown Group Size	Unknown Number of Groups
Equal groups	There are 5 students, and each student has 4 pieces of candy. How many pieces of candy are there in all? $5 \times 4 = ?$	14 pieces of candy are shared equally by 7 students. How many pieces of candy does each student have? $7 \times ? = 14$ Solved by inverse operations $14 \div 7 = ?$	If 18 pieces of candy are to be given out 3 to each student, how many students will get candy? $? \times 3 = 18$ Solved by inverse operations $18 \div 3 = ?$

	Unknown Product	Unknown Factor	Unknown Factor
Arrays	There are 5 rows of students with 3 students in each row. How many students are there? $5 \times 3 = ?$	If 16 students are arranged into 4 equal rows, how many students will be in each row? $4 \times ? = 16$ Solved by inverse operations $16 \div 4 = ?$	If 24 students are arranged into an array with 6 columns, how many rows are there? $? \times 6 = 24$ Solved by inverse operations $24 \div 6 = ?$

	Larger Unknown	Smaller Unknown	Multiplier Unknown
Comparing	A small popcorn costs $1.50. A large popcorn costs 3 times as much as a small popcorn. How much does a large popcorn cost? $1.50 \times 3 = ?$	A large soda costs $6 and that is 2 times as much as a small soda costs. How much does a small soda cost? $2 \times ? = 6$ Solved by inverse operations $6 \div 2 = ?$	A large pretzel costs $3 and a small pretzel costs $2. How many times as much does the large pretzel cost as the small pretzel? $? \times 2 = 3$ Solved by inverse operations $3 \div 2 = ?$

Fraction Operations

A **fraction** is a part of something that is whole. Items such as apples can be cut into parts to help visualize fractions. If an apple is cut into 2 equal parts, each part represents ½ of the apple. If each half is cut into two parts, the apple now is cut into quarters. Each piece now represents ¼ of the apple. In this example, each part is equal because they all have the same size. Geometric shapes, such as circles and squares, can also be utilized to help visualize the idea of fractions. For example, a circle can be drawn and divided into 6 equal parts:

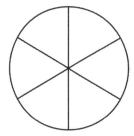

Shading can be used to represent parts of the circle that can be translated into fractions. The top of the fraction, the **numerator**, can represent how many segments are shaded. The bottom of the fraction, the **denominator**, can represent the number of segments that the circle is broken into. A pie is a good

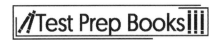

analogy to use in this example. If one piece of the circle is shaded, or one piece of pie is cut out, $^1/_6$ of the object is being referred to. An apple, a pie, or a circle can be utilized in order to compare simple fractions. For example, showing that ½ is larger than ¼ and that ¼ is smaller than $^1/_3$ can be accomplished through shading. A **unit fraction** is a fraction in which the numerator is 1, and the denominator is a positive whole number. It represents one part of a whole—one piece of pie.

Imagine that an apple pie has been baked for a holiday party, and the full pie has eight slices. After the party, there are five slices left. How could the amount of the pie that remains be expressed as a fraction? The numerator is 5 since there are 5 pieces left, and the denominator is 8 since there were eight total slices in the whole pie. Thus, expressed as a fraction, the leftover pie totals $\frac{5}{8}$ of the original amount.

Fractions come in three different varieties: proper fractions, improper fractions, and mixed numbers. **Proper fractions** have a numerator less than the denominator, such as $\frac{3}{8}$, but **improper fractions** have a numerator greater than the denominator, such as $\frac{15}{8}$. **Mixed numbers** combine a whole number with a proper fraction, such as $3\frac{1}{2}$. Any mixed number can be written as an improper fraction by multiplying the integer by the denominator, adding the product to the value of the numerator, and dividing the sum by the original denominator. For example:

$$3\frac{1}{2} = \frac{3 \times 2 + 1}{2} = \frac{7}{2}$$

Whole numbers can also be converted into fractions by placing the whole number as the numerator and making the denominator 1. For example, $3 = \frac{3}{1}$.

The bar in a fraction represents division. Therefore $^6/_5$ is the same as $6 \div 5$. In order to rewrite it as a mixed number, division is performed to obtain $6 \div 5 = 1\ R1$. The remainder is then converted into fraction form. The actual remainder becomes the numerator of a fraction, and the divisor becomes the denominator. Therefore $1\ R1$ is written as $1\frac{1}{5}$, a mixed number. A mixed number can also decompose into the addition of a whole number and a fraction. For example,

$$1\frac{1}{5} = 1 + \frac{1}{5} \text{ and } 4\frac{5}{6} = 4 + \frac{1}{6} + \frac{1}{6} + \frac{1}{6} + \frac{1}{6} + \frac{1}{6}$$

Every fraction can be built from a combination of unit fractions.

One of the most fundamental concepts of fractions is their ability to be manipulated by multiplication or division. This is possible since $\frac{n}{n} = 1$ for any non-zero integer. As a result, multiplying or dividing by $\frac{n}{n}$ will not alter the original fraction since any number multiplied or divided by 1 doesn't change the value of that number. Fractions of the same value are known as equivalent fractions. For example, $\frac{2}{8}, \frac{25}{100},$ and $\frac{40}{160}$ are equivalent, as they all equal $\frac{1}{4}$.

Like fractions, or **equivalent fractions**, are the terms used to describe these fractions that are made up of different numbers but represent the same quantity. For example, the given fractions are $^4/_8$ and $^3/_6$. If a pie was cut into 8 pieces and 4 pieces were removed, half of the pie would remain. Also, if a pie was split into 6 pieces and 3 pieces were eaten, half of the pie would also remain. Therefore, both of the fractions represent half of a pie. These two fractions are referred to as like fractions. **Unlike fractions**

are fractions that are different and cannot be thought of as representing equal quantities. When working with fractions in mathematical expressions, like fractions should be simplified. Both $^4/_8$ and $^3/_6$ can be simplified into $^1/_2$.

Comparing fractions can be completed through the use of a number line. For example, if $^3/_5$ and $^6/_{10}$ need to be compared, each fraction should be plotted on a number line. To plot $^3/_5$, the area from 0 to 1 should be broken into 5 equal segments, and the fraction represents 3 of them. To plot $^6/_{10}$, the area from 0 to 1 should be broken into 10 equal segments, and the fraction represents 6 of them.

It can be seen that $\frac{3}{5} = \frac{6}{10}$

Like fractions are plotted at the same point on a number line. Unit fractions can also be used to compare fractions. For example, if it is known that

$$\frac{4}{5} > \frac{1}{2}$$

and

$$\frac{1}{2} > \frac{4}{10}$$

then it is also known that

$$\frac{4}{5} > \frac{4}{10}$$

Also, converting improper fractions to mixed numbers can be helpful in comparing fractions because the whole number portion of the number is more visible.

Adding and subtracting mixed numbers and fractions can be completed by decomposing fractions into a sum of whole numbers and unit fractions. For example, the given problem is

$$5\frac{3}{7} + 2\frac{1}{7}$$

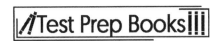

Decomposing into

$$5 + \frac{1}{7} + \frac{1}{7} + \frac{1}{7} + 2 + \frac{1}{7}$$

This shows that the whole numbers can be added separately from the unit fractions. The answer is:

$$5 + 2 + \frac{1}{7} + \frac{1}{7} + \frac{1}{7} + \frac{1}{7} = 7 + \frac{4}{7} = 7\frac{4}{7}$$

Although many equivalent fractions exist, they are easier to compare and interpret when reduced or simplified. The numerator and denominator of a simple fraction will have no factors in common other than 1. When reducing or simplifying fractions, divide the numerator and denominator by the greatest common factor. A simple strategy is to divide the numerator and denominator by low numbers, like 2, 3, or 5 until arriving at a simple fraction, but the same thing could be achieved by determining the greatest common factor for both the numerator and denominator and dividing each by it. Using the first method is preferable when both the numerator and denominator are even, end in 5, or are obviously a multiple of another number. However, if no numbers seem to work, it will be necessary to factor the numerator and denominator to find the GCF. Let's look at examples:

1) Simplify the fraction $\frac{6}{8}$:

Dividing the numerator and denominator by 2 results in $\frac{3}{4}$, which is a simple fraction.

2) Simplify the fraction $\frac{12}{36}$:

Dividing the numerator and denominator by 2 leaves $\frac{6}{18}$. This isn't a simple fraction, as both the numerator and denominator have factors in common. Dividing each by 3 results in $\frac{2}{6}$, but this can be further simplified by dividing by 2 to get $\frac{1}{3}$. This is the simplest fraction, as the numerator is 1. In cases like this, multiple division operations can be avoided by determining the greatest common factor between the numerator and denominator.

3) Simplify the fraction $\frac{18}{54}$ by dividing by the greatest common factor:

First, determine the factors for the numerator and denominator. The factors of 18 are 1, 2, 3, 6, 9, and 18. The factors of 54 are 1, 2, 3, 6, 9, 18, 27, and 54. Thus, the greatest common factor is 18. Dividing $\frac{18}{54}$ by 18 leaves $\frac{1}{3}$, which is the simplest fraction. This method takes slightly more work, but it definitively arrives at the simplest fraction.

Adding and Subtracting Fractions

Adding and subtracting fractions that have the same denominators involves adding or subtracting the numerators. The denominator will stay the same. Therefore, the decomposition process can be made simpler, and the fractions do not have to be broken into unit fractions.

For example, the given problem is:

$$4\frac{7}{8} - 2\frac{6}{8}$$

The answer is found by adding the answers to both

$$4 - 2 \text{ and } \frac{7}{8} - \frac{6}{8}$$

$$2 + \frac{1}{8} = 2\frac{1}{8}$$

A common mistake would be to add the denominators so that

$$\frac{1}{4} + \frac{1}{4} = \frac{1}{8} \text{ or } \frac{2}{8}$$

However, conceptually, it is known that two quarters make a half, so neither one of these are correct.

If two fractions have different denominators, equivalent fractions must be used to add or subtract them. The fractions must be converted into fractions that have common denominators. A **least common denominator** or the product of the two denominators can be used as the common denominator. For example, in the problem $\frac{5}{6} + \frac{2}{3}$, both 6, which is the least common denominator, and 18, which is the product of the denominators, can be used. In order to use 6, $\frac{2}{3}$ must be converted to sixths. A number line can be used to show the equivalent fraction is $\frac{4}{6}$. What happens is that $\frac{2}{3}$ is multiplied times a fractional form of 1 to obtain a denominator of 6. Hence:

$$\frac{2}{3} \times \frac{2}{2} = \frac{4}{6}$$

Therefore, the problem is now $\frac{5}{6} + \frac{4}{6} = \frac{9}{6}$, which can be simplified into $\frac{3}{2}$. In order to use 18, both fractions must be converted into having 18 as their denominator. $\frac{5}{6}$ would have to be multiplied times $\frac{3}{3}$, and $\frac{2}{3}$ would need to be multiplied times $\frac{6}{6}$. The addition problem would be $\frac{15}{18} + \frac{12}{18} = \frac{27}{18}$, which reduces into $\frac{3}{2}$.

It is always possible to find a common denominator by multiplying the denominators. However, when the denominators are large numbers, this method is unwieldy, especially if the answer must be provided in its simplest form. Thus, it's beneficial to find the **least common denominator** of the fractions—the least common denominator is incidentally also the **least common multiple**.

Once equivalent fractions have been found with common denominators, simply add or subtract the numerators to arrive at the answer:

1) $\frac{1}{2} + \frac{3}{4} = \frac{2}{4} + \frac{3}{4} = \frac{5}{4}$

2) $\frac{3}{12} + \frac{11}{20} = \frac{15}{60} + \frac{33}{60} = \frac{48}{60} = \frac{4}{5}$

3) $\frac{7}{9} - \frac{4}{15} = \frac{35}{45} - \frac{12}{45} = \frac{23}{45}$

4) $\frac{5}{6} - \frac{7}{18} = \frac{15}{18} - \frac{7}{18} = \frac{8}{18} = \frac{4}{9}$

Multiplying and Dividing Fractions

Of the four basic operations that can be performed on fractions, the one that involves the least amount of work is multiplication. To multiply two fractions, simply multiply the numerators together, multiply the denominators together, and place the products of each as a fraction. Whole numbers and mixed numbers can also be expressed as a fraction, as described above, to multiply with a fraction.

Because multiplication is commutative, multiplying a fraction times a whole number is the same as multiplying a whole number times a fraction. The problem involves adding a fraction a specific number of times. The problem $3 \times \frac{1}{4}$ can be translated into adding the unit fraction 3 times:

$$\frac{1}{4} + \frac{1}{4} + \frac{1}{4} = \frac{3}{4}$$

In the problem $4 \times \frac{2}{5}$, the fraction can be decomposed into $\frac{1}{5} + \frac{1}{5}$ and then added 4 times to obtain $\frac{8}{5}$. Also, both of these answers can be found by just multiplying the whole number times the numerator of the fraction being multiplied.

The whole numbers can be written in fraction form as:

$$\frac{3}{1} \times \frac{1}{4} = \frac{3}{4}$$

$$\frac{4}{1} \times \frac{2}{5} = \frac{8}{5}$$

Multiplying a fraction times a fraction involves multiplying the numerators together separately and the denominators together separately. For example,

$$\frac{3}{8} \times \frac{2}{3} = \frac{3 \times 2}{8 \times 3} = \frac{6}{24}$$

This can then be reduced to $^{1}/_{4}$.

Dividing a fraction by a fraction is actually a multiplication problem. It involves flipping the divisor and then multiplying normally. For example,

$$\frac{22}{5} \div \frac{1}{2} = \frac{22}{5} \times \frac{2}{1} = \frac{44}{5}$$

The same procedure can be implemented for division problems involving fractions and whole numbers. The whole number can be rewritten as a fraction over a denominator of 1, and then division can be completed.

A common denominator approach can also be used in dividing fractions. Considering the same problem, $\frac{22}{5} \div \frac{1}{2}$, a common denominator between the two fractions is 10. $\frac{22}{5}$ would be rewritten as $\frac{22}{5} \times \frac{2}{2} = \frac{44}{10}$, and $\frac{1}{2}$ would be rewritten as:

$$\frac{1}{2} \times \frac{5}{5} = \frac{5}{10}$$

Dividing both numbers straight across results in:

$$\frac{44}{10} \div \frac{5}{10} = \frac{44/5}{10/10} = \frac{44/5}{1} = 44/5$$

Many real-world problems will involve the use of fractions. Key words include actual fraction values, such as half, quarter, third, fourth, etc. The best approach to solving word problems involving fractions is to draw a picture or diagram that represents the scenario being discussed, while deciding which type of operation is necessary in order to solve the problem. A phrase such as "one fourth of 60 pounds of coal" creates a scenario in which multiplication should be used, and the mathematical form of the phrase is $\frac{1}{4} \times 60$.

Decimal Operations

The **decimal system** is a way of writing out numbers that uses ten different numerals: 0, 1, 2, 3, 4, 5, 6, 7, 8, and 9. This is also called a "base ten" or "base 10" system. Other bases are also used. For example, computers work with a base of 2. This means they only use the numerals 0 and 1.

The **decimal place** denotes how far to the right of the decimal point a numeral is. The first digit to the right of the decimal point is in the **tenths** place. The next is the **hundredths**. The third is the *thousandths*.

So, 3.142 has a 1 in the tenths place, a 4 in the hundredths place, and a 2 in the thousandths place.

The **decimal point** is a period used to separate the *ones* place from the *tenths* place when writing out a number as a decimal.

A **decimal number** is a number written out with a decimal point instead of as a fraction, for example, 1.25 instead of $\frac{5}{4}$. Depending on the situation, it can sometimes be easier to work with fractions and sometimes easier to work with decimal numbers.

A decimal number is **terminating** if it stops at some point. It is called **repeating** if it never stops but repeats a pattern over and over. It is important to note that every rational number can be written as a terminating decimal or as a repeating decimal.

Addition with Decimals

To add decimal numbers, each number needs to be lined up by the decimal point in vertical columns. For each number being added, the zeros to the right of the last number need to be filled in so that each of the numbers has the same number of places to the right of the decimal. Then, the columns can be added together. Here is an example of 2.45 + 1.3 + 8.891 written in column form:

$$2.450$$

$$1.300$$

$$+ \ 8.891$$

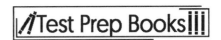

Zeros have been added in the columns so that each number has the same number of places to the right of the decimal.

Added together, the correct answer is 12.641:

$$2.450$$

$$1.300$$

$$+\ 8.891$$

$$12.641$$

Subtraction with Decimals

Subtracting decimal numbers is the same process as adding decimals. Here is $7.89 - 4.235$ written in column form:

$$7.890$$

$$-\ 4.235$$

$$3.655$$

A zero has been added in the column so that each number has the same number of places to the right of the decimal.

Multiplication with Decimals

The simplest way to multiply decimals is to calculate the product as if the decimals are not there, then count the number of decimal places in the original problem. Use that total to place the decimal the same number of places over in your answer, counting from right to left. For example, 0.5 x 1.25 can be rewritten and multiplied as 5 x 125, which equals 625. Then the decimal is added three places from the right for .625.

The final answer will have the same number of decimal *points* as the total number of decimal *places* in the problem. The first number has one decimal place, and the second number has two decimal places. Therefore, the final answer will contain three decimal places:

$$0.5 \times 1.25 = 0.625$$

Division with Decimals

Dividing a decimal by a whole number entails using long division first by ignoring the decimal point. Then, the decimal point is moved the number of places given in the problem.

For example, $6.8 \div 4$ can be rewritten as $68 \div 4$, which is 17. There is one non-zero integer to the right of the decimal point, so the final solution would have one decimal place to the right of the solution. In this case, the solution is 1.7.

Dividing a decimal by another decimal requires changing the divisor to a whole number by moving its decimal point. The decimal place of the dividend should be moved by the same number of places as the divisor. Then, the problem is the same as dividing a decimal by a whole number.

For example, $5.72 \div 1.1$ has a divisor with one decimal point in the denominator. The expression can be rewritten as $57.2 \div 11$ by moving each number one decimal place to the right to eliminate the decimal. The long division can be completed as $572 \div 11$ with a result of 52. Since there is one non-zero integer to the right of the decimal point in the problem, the final solution is 5.2.

In another example, $8 \div 0.16$ has a divisor with two decimal points in the denominator. The expression can be rewritten as $800 \div 16$ by moving each number two decimal places to the right to eliminate the decimal in the divisor. The long division can be completed with a result of 50.

Estimation and Rounding with Decimals

Prior to performing operations and calculating the answer to a problem involving addition, subtraction, multiplication, or division, it is helpful to estimate the result. Doing so will enable the test taker to determine whether his or her computed answer is logical within the context of a given problem and prevent careless errors. For example, it is unfortunately common under the pressure of a testing situation for test takers to inadvertently perform the incorrect operation or make a simple calculation error on an otherwise easy math problem. By quickly estimating the answer by eyeballing the numbers, rounding if needed, and performing some simple mental math, test takers can establish an approximate expected outcome before calculating the specific answer. The derived result after computation can then be evaluated by its nearness to the expected answer. This is performed by approximating given values to perform mental math. Numbers should be rounded to the nearest value possible to check the initial results.

As mentioned, sometimes when performing operations such as multiplying numbers, the result can be estimated by rounding. For example, to estimate the value of 11.2×2.01, each number can be rounded to the nearest integer. This will yield a result of 22.

Rounding numbers helps with estimation because it changes the given number to a simpler, although less accurate, number than the exact given number. Rounding allows for easier calculations, which estimate the results of using the exact given number. The accuracy of the estimate and ease of use depends on the place value to which the number is rounded. First, the place value is specified. Then, the digit to its right is looked at. For example, if rounding to the nearest hundreds place, the digit in the tens place is used. If it is a zero, one, 2, 3, or 4, the digit being rounded to is left alone. If it is a 5, 6, 7, 8 or 9, the digit being rounded to is increased by one. All other digits before the decimal point are then changed to zeros, and the digits in decimal places are dropped. If a decimal place is being rounded to, all digits that come after are just dropped.

For example, if 845,231.45 was to be rounded to the nearest thousands place, the answer would be 845,000. The 5 would remain the same due to the 2 in the hundreds place. Also, if 4.567 were to be rounded to the nearest tenths place, the answer would be 4.6. The 5 increased to 6 due to the 6 in the hundredths place, and the rest of the decimal is dropped.

Percent

Percent Problems

Think of percentages as fractions with a denominator of 100. In fact, percentage means "per hundred." The basic percent equation is the following:

$$\frac{is}{of} = \frac{\%}{100}$$

The placement of numbers in the equation depends on what the question asks.

Example 1
Find 40% of 80.

Basically, the problem is asking, "What is 40% of 80?" The 40% is the percent, and 80 is the number to find the percent "of." The equation is:

$$\frac{x}{80} = \frac{40}{100}$$

After cross-multiplying, the problem becomes 100x = 80(40). Solving for x gives the answer: x = 32.

Example 2
What percent of 100 is 20?

The 20 fills in the "is" portion, while 100 fills in the "of." The question asks for the percent, so that will be x, the unknown. The following equation is set up:

$$\frac{20}{100} = \frac{x}{100}$$

Cross-multiplying yields the equation 100x = 20(100). Solving for x gives the answer of 20%.

Example 3
30% of what number is 30?

The following equation uses the clues and numbers in the problem:

$$\frac{30}{x} = \frac{30}{100}$$

Cross-multiplying results in the equation 30(100) = 30x. Solving for x gives the answer x = 100.

Relationships in Numerical Data

In some cases, it is useful to compare numerical data and determine the relationship between values. One of the best ways to mathematically compare two values is to compute the percentage difference between the two values. For example, consider a given music shop that had a net profit of $120,000 in the first year of operation and $185,000 over the second year. Rather than simply finding the net difference between the two years (using subtraction), the business owner may want to know by what percentage his profit increased; in other words, how much his profit in the second year increased relative to his first year. In such cases, the percentage change is desired. The following sections provide some guidance for this process.

Percent Increase/Decrease

Problems dealing with percentages may involve an original value, a change in that value, and a percentage change. A problem will provide two pieces of information and ask to find the third. To do so, this formula is used:

$$\frac{change}{original\ value} \text{x } 100 = \text{percent change}$$

Here's a sample problem:

> Attendance at a baseball stadium has dropped 16% from last year. Last year's average attendance was 40,000. What is this year's average attendance?

Using the formula and information, the change is unknown (x), the original value is 40,000, and the percent change is 16%. The formula can be written as:

$$\frac{x}{40,000} \text{x } 100 = 16$$

When solving for x, it is determined the change was 6,400. The problem asked for this year's average attendance, so to calculate, the change (6,400) is subtracted from last year's attendance (40,000) to determine this year's average attendance is 33,600.

Percent More Than/Less Than

Percentage problems may give a value and what percent that given value is more than or less than an original unknown value. Here's a sample problem:

> A store advertises that all its merchandise has been reduced by 25%. The new price of a pair of shoes is $60. What was the original price?

This problem can be solved by writing a proportion. Two ratios should be written comparing the cost and the percent of the original cost. The new cost is 75% of the original cost (100% - 25%); and the original cost is 100% of the original cost. The unknown original cost can be represented by x. The proportion would be set up as:

$$\frac{60}{75} = \frac{x}{100}$$

Solving the proportion, it is determined the original cost was $80.

Real-World Rate, Percent, and Measurement Problems

A **ratio** compares the size of one group to the size of another. For example, there may be a room with 4 tables and 24 chairs. The ratio of tables to chairs is 4: 24. Such ratios behave like fractions in that both sides of the ratio by the same number can be multiplied or divided. Thus, the ratio 4:24 is the same as the ratio 2:12 and 1:6.

One quantity is **proportional** to another quantity if the first quantity is always some multiple of the second. For instance, the distance travelled in five hours is always five times to the speed as travelled. The distance is proportional to speed in this case.

One quantity is **inversely proportional** to another quantity if the first quantity is equal to some number divided by the second quantity. The time it takes to travel one hundred miles will be given by 100 divided by the speed travelled. The time is inversely proportional to the speed.

When dealing with word problems, there is no fixed series of steps to follow, but there are some general guidelines to use. It is important that the quantity to be found is identified. Then, it can be determined how the given values can be used and manipulated to find the final answer.

Example 1
Jana wants to travel to visit Alice, who lives one hundred and fifty miles away. If she can drive at fifty miles per hour, how long will her trip take?

The quantity to find is the *time* of the trip. The time of a trip is given by the distance to travel divided by the speed to be traveled. The problem determines that the distance is one hundred and fifty miles, while the speed is fifty miles per hour. Thus, 150 divided by 50 is $150 \div 50 = 3$. Because *miles* and *miles per hour* are the units being divided, the miles cancel out. The result is 3 hours.

Example 2
Bernard wishes to paint a wall that measures twenty feet wide by eight feet high. It costs ten cents to paint one square foot. How much money will Bernard need for paint?

The final quantity to compute is the *cost* to paint the wall. This will be ten cents ($0.10) for each square foot of area needed to paint. The area to be painted is unknown, but the dimensions of the wall are given; thus, it can be calculated.

The dimensions of the wall are 20 feet wide and 8 feet high. Since the area of a rectangle is length multiplied by width, the area of the wall is 8 x 20 = 160 square feet. Multiplying 0.1 x 160 yields $16 as the cost of the paint.

Number Comparison and Equivalents

The Position of Numbers Relative to Each Other

Place Value of a Digit
Numbers count in groups of 10. That number is the same throughout the set of natural numbers and whole numbers. It is referred to as working within a base 10 numeration system. Only the numbers from zero to 9 are used to represent any number. The foundation for doing this involves **place value**. Numbers are written side by side. This is to show the amount in each place value.

For place value, let's look at how the number 10 is different from zero to 9. It has two digits instead of just one. The one is in the tens' place, and the zero is in the ones' place. Therefore, there is one group of tens and zero ones. 11 has one 10 and one 1. The introduction of numbers from 11 to 19 should be the next step. Each value within this range of numbers consists of one group of 10 and a specific number of leftover ones. Counting by tens can be practiced once the tens column is understood. This process consists of increasing the number in the tens place by one. For example, counting by 10 starting at 17 would result in the next four values being 27, 37, 47, and 57.

A place value chart can be used for understanding and learning about numbers that have more digits. Here is an example of a place value chart:

	MILLIONS			THOUSANDS			ONES			.	DECIMALS		
billions	hundred millions	ten millions	millions	hundred thousands	ten thousands	thousands	hundreds	tens	ones		tenths	hundredths	thousandths

In the number 1,234, there are 4 ones and 3 tens. The 2 is in the hundreds' place, and the one is in the thousands' place. Note that each group of three digits is separated by a comma. The 2 has a value that is 10 times greater than the 3. Every place to the left has a value 10 times greater than the place to its right. Also, each group of three digits is also known as a *period*. 234 is in the ones' period.

The number 1,234 can be written out as *one-thousand, two hundred thirty-four*. The process of writing out numbers is known as the *decimal system*. It is also based on groups of 10. The place value chart is a helpful tool in using this system. In order to write out a number, it always starts with the digit(s) in the highest period. For example, in the number 23,815,467, the 23 is in highest place and is in the millions' period. The number is read *twenty-three million, eight hundred fifteen thousand, four hundred sixty-seven*. Each period is written separately through the use of commas. Also, no "ands" are used within the number. Another way to think about the number 23,815,467 is through the use of an addition problem. For example:

$$23,815,467 = 20,000,000 + 3,000,000 + 800,000 + 10,000 + 5,000 + 400 + 60 + 7$$

This expression is known as *expanded form*. The actual number 23,815,467 is known as being in **standard form**.

In order to compare whole numbers with many digits, place value can be used. In each number to be compared, it is necessary to find the highest place value in which the numbers differ and to compare the value within that place value. For example, $4,523,345 < 4,532,456$ because of the values in the ten thousands place. A similar process can be used for decimals. However, number lines can also be used. Tick marks can be placed within two whole numbers on the number line that represent tenths, hundredths, etc. Each number being compared can then be plotted. The value farthest to the right on the number line is the largest.

Comparing, Classifying, and Ordering Real Numbers

Rational numbers are any number that can be written as a fraction or ratio. Within the set of rational numbers, several subsets exist that are referenced throughout the mathematics topics. **Counting numbers** are the first numbers learned as a child. Counting numbers consist of 1,2,3,4, and so on. **Whole numbers** include all counting numbers and zero (0,1,2,3,4,...). **Integers** include counting numbers, their opposites, and zero (...,-3,-2,-1,0,1,2,3,...). **Rational numbers** are inclusive of integers, fractions, and decimals that terminate, or end (1.7, 0.04213) or repeat (0.136$\overline{5}$).

A **number line** typically consists of integers (...3,2,1,0,-1,-2,-3...), and is used to visually represent the value of a rational number. Each rational number has a distinct position on the line determined by comparing its value with the displayed values on the line. For example, if plotting -1.5 on the number line below, it is necessary to recognize that the value of -1.5 is .5 less than -1 and .5 greater than -2. Therefore, -1.5 is plotted halfway between -1 and -2.

The number system that is used consists of only ten different digits or characters. However, this system is used to represent an infinite number of values. As mentioned, the **place value system** makes this infinite number of values possible. The position in which a digit is written corresponds to a given value. Starting from the decimal point (which is implied, if not physically present), each subsequent place value to the left represents a value greater than the one before it. Conversely, starting from the decimal point, each subsequent place value to the right represents a value less than the one before it.

In accordance with the **base-10 system**, the value of a digit increases by a factor of ten each place it moves to the left. For example, consider the number 7. Moving the digit one place to the left (70), increases its value by a factor of 10 ($7 \times 10 = 70$).

Moving the digit two places to the left (700) increases its value by a factor of 10 twice ($7 \times 10 \times 10 = 700$). Moving the digit three places to the left (7,000) increases its value by a factor of 10 three times ($7 \times 10 \times 10 \times 10 = 7,000$), and so on.

Conversely, the value of a digit decreases by a factor of ten each place it moves to the right. (Note that multiplying by $\frac{1}{10}$ is equivalent to dividing by 10). For example, consider the number 40. Moving the digit one place to the right (4) decreases its value by a factor of 10 ($40 \div 10 = 4$). Moving the digit two places to the right (0.4), decreases its value by a factor of 10 twice:

$$(40 \div 10 \div 10 = 0.4)$$

or

$$(40 \times \frac{1}{10} \times \frac{1}{10} = 0.4)$$

Moving the digit three places to the right (0.04) decreases its value by a factor of 10 three times:

$$(40 \div 10 \div 10 \div 10 = 0.04)$$

or

$$(40 \times \frac{1}{10} \times \frac{1}{10} \times \frac{1}{10} = 0.04), \text{ and so on.}$$

Ordering Numbers

A common question type asks to order rational numbers from least to greatest or greatest to least. The numbers will come in a variety of formats, including decimals, percentages, roots, fractions, and whole numbers. These questions test for knowledge of different types of numbers and the ability to determine their respective values.

Before discussing ordering all numbers, let's start with decimals.

To compare decimals and order them by their value, utilize a method similar to that of ordering large numbers.

The main difference is where the comparison will start. Assuming that any numbers to left of the decimal point are equal, the next numbers to be compared are those immediately to the right of the decimal point. If those are equal, then move on to compare the values in the next decimal place to the right.

For example:

Which number is greater, 12.35 or 12.38?

Check that the values to the left of the decimal point are equal:

12 = 12

Next, compare the values of the decimal place to the right of the decimal:

12.3 = 12.3

Those are also equal in value.

Finally, compare the value of the numbers in the next decimal place to the right on both numbers:

12.3**5** and 12.3**8**

Here the 5 is less than the 8, so the final way to express this inequality is:

12.35 < 12.38

Comparing decimals is regularly exemplified with money because the "cents" portion of money ends in the hundredths place. When paying for gasoline or meals in restaurants, and even in bank accounts, if enough errors are made when calculating numbers to the hundredths place, they can add up to dollars and larger amounts of money over time.

Now that decimal ordering has been explained, let's expand and consider all real numbers. Whether the question asks to order the numbers from greatest to least or least to greatest, the crux of the question is the same—convert the numbers into a common format. Generally, it's easiest to write the numbers as whole numbers and decimals so they can be placed on a number line. Follow these examples to understand this strategy.

1) Order the following rational numbers from greatest to least:

$$\sqrt{36}, 0.65, 78\%, \frac{3}{4}, 7, 90\%, \frac{5}{2}$$

Of the seven numbers, the whole number (7) and decimal (0.65) are already in an accessible form, so concentrate on the other five.

First, the square root of 36 equals 6. (If the test asks for the root of a non-perfect root, determine which two whole numbers the root lies between.) Next, convert the percentages to decimals. A percentage means "per hundred," so this conversion requires moving the decimal point two places to the left, leaving 0.78 and 0.9.

Lastly, evaluate the fractions:

$$\frac{3}{4} = \frac{75}{100} = 0.75 \; ; \frac{5}{2} = 2\frac{1}{2} = 2.5$$

Now, the only step left is to list the numbers in the request order:

$$7, \sqrt{36}, \frac{5}{2}, 90\%, 78\%, \frac{3}{4}, 0.65$$

2) Order the following rational numbers from least to greatest:

$$2.5, \sqrt{9}, -10.5, 0.853, 175\%, \sqrt{4}, \frac{4}{5}$$

$$\sqrt{9} = 3$$

$$175\% = 1.75$$

$$\sqrt{4} = 2$$

$$\frac{4}{5} = 0.8$$

From least to greatest, the answer is:

$$-10.5, \frac{4}{5}, 0.853, 175\%, \sqrt{4}, 2.5, \sqrt{9}$$

Expressing Numeric Relationships

If a question asks to give words to a mathematical expression and says "equals," then an = sign must be included in the answer. Similarly, "less than or equal to" is expressed by the inequality symbol ≤, and "greater than or equal" to is expressed as ≥. Furthermore, "less than" is represented by <, and "greater than" is expressed by >.

Equations use the equals sign because the numeric expressions on either side of the symbol (=) are equivalent. In contrast, inequalities compare values or expressions that are unequal. Although not always true, linear equations that include a variable often have just one value for the variable that makes the statement true. Linear inequalities generally have an infinite number of values that make the statement true.

Inequalities are a concise mathematical way to express the relationship between unequal values. More specifically, they describe in what way the values are unequal. A value could be greater than ($>$); less than ($<$); greater than or equal to ($\geq$); or less than or equal to ($\leq$) another value. The statement "five times a number added to forty is more than sixty-five" can be expressed as $5x + 40 > 65$. Common words and phrases that express inequalities are:

Symbol	Phrase
$<$	is under, is below, smaller than, beneath
$>$	is above, is over, bigger than, exceeds
$\leq$	no more than, at most, maximum
$\geq$	no less than, at least, minimum

Conversions

Decimals and Percentages

Since a percentage is based on "per hundred," decimals and percentages can be converted by multiplying or dividing by 100. Practically speaking, this always amounts to moving the decimal point two places to the right or left, depending on the conversion. To convert a percentage to a decimal, move the decimal point two places to the left and remove the % sign. To convert a decimal to a percentage, move the decimal point two places to the right and add a "%" sign. Here are some examples:

65% = 0.65
0.33 = 33%
0.215 = 21.5%
99.99% = 0.9999
500% = 5.00
7.55 = 755%

Fractions and Percentages

Remember that a percentage is a number per one hundred. So a percentage can be converted to a fraction by making the number in the percentage the numerator and putting 100 as the denominator:

$$43\% = \frac{43}{100}$$

$$97\% = \frac{97}{100}$$

Note that the percent symbol (%) kind of looks like a 0, a 1, and another 0. So think of a percentage like 54% as 54 over 100.

To convert a fraction to a percent, follow the same logic. If the fraction happens to have 100 in the denominator, you're in luck. Just take the numerator and add a percent symbol:

$$\frac{28}{100} = 28\%$$

Otherwise, divide the numerator by the denominator to get a decimal:

$$\frac{9}{12} = 0.75$$

Then convert the decimal to a percentage:

$$0.75 = 75\%$$

Another option is to make the denominator equal to 100. Be sure to multiply the numerator and the denominator by the same number. For example:

$$\frac{3}{20} \times \frac{5}{5} = \frac{15}{100}$$

$$\frac{15}{100} = 15\%$$

Changing Fractions to Decimals

To change a fraction into a decimal, divide the denominator into the numerator until there are no remainders. There may be repeating decimals, so rounding is often acceptable. A straight line above the repeating portion denotes that the decimal repeats.

Example: Express 4/5 as a decimal.

Set up the division problem.

$$5\overline{)4}$$

5 does not go into 4, so place the decimal and add a zero.

$$5\overline{)4.0}$$

5 goes into 40 eight times. There is no remainder.

$$\begin{array}{r} 0.8 \\ 5\overline{)4.0} \\ -\ 4.0 \\ \hline 0 \end{array}$$

The solution is 0.8.

Example: Express 33 1/3 as a decimal.

Since the whole portion of the number is known, set it aside to calculate the decimal from the fraction portion.

Set up the division problem.

$$3\overline{)1}$$

3 does not go into 1, so place the decimal and add zeros. 3 goes into 10 three times.

$$\begin{array}{r} 0.3 \\ 3\overline{)1.0} \end{array}$$

This will repeat with a remainder of 1.

$$\begin{array}{r} 0.333 \\ 3\overline{)1.000} \\ -9 \\ \hline 10 \\ -9 \\ \hline 10 \end{array}$$

So, we will place a line over the 3 to denote the repetition. The solution is written $0.\overline{3}$.

Changing Decimals to Fractions

To change decimals to fractions, place the decimal portion of the number, the numerator, over the respective place value, the denominator, then reduce, if possible.

Example: Express 0.25 as a fraction.

This is read as twenty-five hundredths, so put 25 over 100. Then reduce to find the solution.

$$\frac{25}{100} = \frac{1}{4}$$

Example: Express 0.455 as a fraction

This is read as four hundred fifty-five thousandths, so put 455 over 1000. Then reduce to find the solution.

$$\frac{455}{1000} = \frac{91}{200}$$

There are two types of problems that commonly involve percentages. The first is to calculate some percentage of a given quantity, where you convert the percentage to a decimal, and multiply the quantity by that decimal. Secondly, you are given a quantity and told it is a fixed percent of an unknown quantity. In this case, convert to a decimal, then divide the given quantity by that decimal.

Example: What is 30% of 760?

Convert the percent into a useable number. "Of" means to multiply.

$$30\% = 0.30$$

Set up the problem based on the givens, and solve.

$$0.30 \times 760 = 228$$

Example: 8.4 is 20% of what number?

Convert the percent into a useable number.

$$20\% = 0.20$$

The given number is a percent of the answer needed, so divide the given number by this decimal rather than multiplying it.

$$\frac{8.4}{0.20} = 42$$

Distribution of a Quantity into its Fractional Parts

A quantity may be broken into its fractional parts. For example, a toy box holds three types of toys for kids. $\frac{1}{3}$ of the toys are Type A and $\frac{1}{4}$ of the toys are Type B. With that information, how many Type C toys are there?

First, the sum of Type A and Type B must be determined by finding a common denominator to add the fractions. The lowest common multiple is 12, so that is what will be used. The sum is:

$$\frac{1}{3} + \frac{1}{4} = \frac{4}{12} + \frac{3}{12} = \frac{7}{12}$$

This value is subtracted from 1 to find the number of Type C toys. The value is subtracted from 1 because 1 represents a whole. The calculation is:

$$1 - \frac{7}{12} = \frac{12}{12} - \frac{7}{12} = \frac{5}{12}$$

This means that $\frac{5}{12}$ of the toys are Type C. To check the answer, add all fractions together, and the result should be 1.

Practice Questions

1. Which of the following numbers has the greatest value?
   a. 1.4378
   b. 1.07548
   c. 1.43592
   d. 0.89409

2. After a 20% sale discount, Frank purchased a new refrigerator for $850. How much did he save from the original price?
   a. $170
   b. $212.50
   c. $105.75
   d. $200

3. A school has 15 teachers and 20 teaching assistants. They have 200 students. What is the ratio of faculty to students?
   a. 3:20
   b. 4:17
   c. 3:2
   d. 7:40

4. Express the solution to the following problem in decimal form:

$$\frac{3}{5} \times \frac{7}{10} \div \frac{1}{2}$$

   a. 0.042
   b. 84%
   c. 0.84
   d. 0.42

5. If Sarah reads at an average rate of 21 pages in four nights, how long will it take her to read 140 pages?
   a. 6 nights
   b. 26 nights
   c. 8 nights
   d. 27 nights

Answer Explanations

1. A: Compare each numeral after the decimal point to figure out which overall number is greatest. In answers A (1.43785) and C (1.43592), both have the same tenths (4) and hundredths (3). However, the thousandths is greater in answer A (7), so A has the greatest value overall.

2. B: Since $850 is the price *after* a 20% discount, $850 represents 80% of the original price. To determine the original price, set up a proportion with the ratio of the sale price (850) to original price (unknown) equal to the ratio of sale percentage (where x represents the unknown original price):

$$\frac{850}{x} = \frac{80}{100}$$

To solve a proportion, cross multiply the numerators and denominators and set the products equal to each other: $(850)(100) = (80)(x)$. Multiplying each side results in the equation $85,000 = 80x$.

To solve for x, divide both sides by 80: $\frac{85,000}{80} = \frac{80x}{80}$, resulting in $x = 1062.5$. Remember that x represents the original price. Subtracting the sale price from the original price ($1062.50 - 850) indicates that Frank saved $212.50.

3. D: The total faculty is 15 + 20 = 35. So the ratio is 35:200. Then, divide both of these numbers by 5, since 5 is a common factor to both, with a result of 7:40.

4. C: The first step in solving this problem is expressing the result in fraction form. Separate this problem first by solving the division operation of the last two fractions. When dividing one fraction by another, invert or flip the second fraction and then multiply the numerator and denominator.

$$\frac{7}{10} \times \frac{2}{1} = \frac{14}{10}$$

Next, multiply the first fraction with this value:

$$\frac{3}{5} \times \frac{14}{10} = \frac{42}{50}$$

In this instance, you can find the decimal form by converting the fraction into $\frac{x}{100}$, where x is the number from which the final decimal is found. Multiply both the numerator and denominator by 2 to get the fraction as an expression of $\frac{x}{100}$.

$$\frac{42}{50} \times \frac{2}{2} = \frac{84}{100}$$

In decimal form, this would be expressed as 0.84.

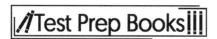

5. D: This problem can be solved by setting up a proportion involving the given information and the unknown value. The proportion is:

$$\frac{21\ pages}{4\ nights} = \frac{140\ pages}{x\ nights}$$

Solving the proportion by cross-multiplying, the equation becomes $21x = 4 * 140$, where $x = 26.67$. Since it is not an exact number of nights, the answer is rounded up to 27 nights. Twenty-six nights would not give Sarah enough time.

Quantitative Reasoning, Algebra, and Statistics

Rational Numbers

All real numbers can be separated into two groups: rational and irrational numbers. **Rational numbers** are any numbers that can be written as a fraction, such as $\frac{1}{3}, \frac{7}{4}$, and -25. Alternatively, **irrational numbers** are those that cannot be written as a fraction, such as numbers with never-ending, non-repeating decimal values. Many irrational numbers result from taking roots, such as $\sqrt{2}$ or $\sqrt{3}$. An irrational number may be written as:

$$34.5684952\ldots$$

The ellipsis (...) represents the line of numbers after the decimal that does not repeat and is never-ending.

When rational and irrational numbers interact, there are different types of number outcomes. For example, when adding or multiplying two rational numbers, the result is a rational number. No matter what two fractions are added or multiplied together, the result can always be written as a fraction. The following expression shows two rational numbers multiplied together:

$$\frac{3}{8} \times \frac{4}{7} = \frac{12}{56}$$

The product of these two fractions is another fraction that can be simplified to $\frac{3}{14}$.

As another interaction, rational numbers added to irrational numbers will always result in irrational numbers. No part of any fraction can be added to a never-ending, non-repeating decimal to make a rational number. The same result is true when multiplying a rational and irrational number. Taking a fractional part of a never-ending, non-repeating decimal will always result in another never-ending, non-repeating decimal. An example of the product of rational and irrational numbers is shown in the following expression: $2 \times \sqrt{7}$.

The last type of interaction concerns two irrational numbers, where the sum or product may be rational or irrational depending on the numbers being used. The following expression shows a rational sum from two irrational numbers:

$$\sqrt{3} + \left(6 - \sqrt{3}\right) = 6$$

The product of two irrational numbers can be rational or irrational. A rational result can be seen in the following expression:

$$\sqrt{2} \times \sqrt{8} = \sqrt{2 \times 8} = \sqrt{16} = 4$$

An irrational result can be seen in the following:

$$\sqrt{3} \times \sqrt{2} = \sqrt{6}$$

Structure of the Number System

The mathematical number system is made up of two general types of numbers: real and complex. **Real numbers** are those that are used in normal settings, while **complex numbers** are those composed of both a real number and an imaginary one. Imaginary numbers are the result of taking the square root of -1, and $\sqrt{-1} = i$.

The real number system is often explained using a Venn diagram similar to the one below. After a number has been labeled as a real number, further classification occurs when considering the other groups in this diagram. If a number is a never-ending, non-repeating decimal, it falls in the irrational category. Otherwise, it is rational. More information on these types of numbers is provided in the previous section. Furthermore, if a number does not have a fractional part, it is classified as an integer, such as -2, 75, or zero. Whole numbers are an even smaller group that only includes positive integers and zero. The last group of natural numbers is made up of only positive integers, such as 2, 56, or 12.

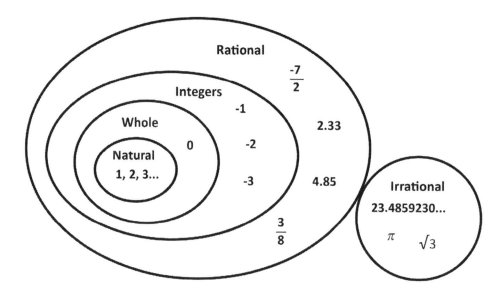

Real numbers can be compared and ordered using the number line. If a number falls to the left on the real number line, it is less than a number on the right. For example, $-2 < 5$ because -2 falls to the left of zero, and 5 falls to the right. Numbers to the left of zero are negative while those to the right are positive.

Complex numbers are made up of the sum of a real number and an imaginary number. Some examples of complex numbers include $6 + 2i$, $5 - 7i$, and $-3 + 12i$. Adding and subtracting complex numbers is similar to collecting like terms. The real numbers are added together, and the imaginary numbers are added together. For example, if the problem asks to simplify the expression $6 + 2i - 3 + 7i$, the 6 and (-3) are combined to make 3, and the $2i$ and $7i$ combine to make $9i$. Multiplying and dividing complex numbers is similar to working with exponents.

One rule to remember when multiplying is that $i \times i = -1$. For example, if a problem asks to simplify the expression $4i(3 + 7i)$, the $4i$ should be distributed throughout the 3 and the $7i$. This leaves the final expression $12i - 28$. The 28 is negative because $i \times i$ results in a negative number. The last type of operation to consider with complex numbers is the conjugate. The **conjugate** of a complex number is a technique used to change the complex number into a real number. For example, the conjugate of $4 - 3i$

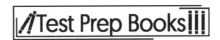

is $4 + 3i$. Multiplying $(4 - 3i)(4 + 3i)$ results in $16 + 12i - 12i + 9$, which has a final answer of $16 + 9 = 25$.

The order of operations—PEMDAS—simplifies longer expressions with real or imaginary numbers. Each operation is listed in the order of how they should be completed in a problem containing more than one operation. Parenthesis can also mean grouping symbols, such as brackets and absolute value. Then, exponents are calculated. Multiplication and division should be completed from left to right, and addition and subtraction should be completed from left to right.

Simplification of another type of expression occurs when radicals are involved. Root is another word for radical. For example, the following expression is a radical that can be simplified: $\sqrt{24x^2}$. First, the number must be factored out to the highest perfect square. Any perfect square can be taken out of a radical. Twenty-four can be factored into 4 and 6, and 4 can be taken out of the radical. $\sqrt{4} = 2$ can be taken out, and 6 stays underneath. If $x > 0$, x can be taken out of the radical because it is a perfect square. The simplified radical is $2x\sqrt{6}$. An approximation can be found using a calculator.

There are also properties of numbers that are true for certain operations. The **commutative** property allows the order of the terms in an expression to change while keeping the same final answer. Both addition and multiplication can be completed in any order and still obtain the same result. However, order does matter in subtraction and division. The **associative** property allows any terms to be "associated" by parenthesis and retain the same final answer. For example:

$$(4 + 3) + 5 = 4 + (3 + 5)$$

Both addition and multiplication are associative; however, subtraction and division do not hold this property. The **distributive** property states that:

$$a(b + c) = ab + ac$$

It is a property that involves both addition and multiplication, and the a is distributed onto each term inside the parentheses.

Absolute Value

The distance that a number is from zero is called its **absolute value**. For example, both 3 and -3 are three spaces from zero. Thus, both -3 and 3 have an absolute value of 3 since they're both three spaces away from zero.

An absolute number is written by placing | | around the number. So, |3| and |−3| both equal 3, as that's their common absolute value.

Ratio and Proportional Relationships

Ratios and Proportional Relationships

Ratios are used to show the relationship between two quantities. The ratio of oranges to apples in the grocery store may be 3 to 2. That means that for every 3 oranges, there are 2 apples. This comparison can be expanded to represent the actual number of oranges and apples, such as 36 oranges to 24 apples. Another example may be the number of boys to girls in a math class. If the ratio of boys to girls is given as 2 to 5, that means there are 2 boys to every 5 girls in the class. Ratios can also be compared if the units in each ratio are the same. The ratio of boys to girls in the math class can be compared to the ratio of boys to girls in a science class by stating which ratio is higher and which is lower.

Rates are used to compare two quantities with different units. **Unit rates** are the simplest form of rate. With unit rates, the denominator in the comparison of two units is one. For example, if someone can type at a rate of 1000 words in 5 minutes, then his or her unit rate for typing is $\frac{1000}{5} = 200$ words in one minute or 200 words per minute. Any rate can be converted into a unit rate by dividing to make the denominator one. 1000 words in 5 minutes has been converted into the unit rate of 200 words per minute.

Ratios and rates can be used together to convert rates into different units. For example, if someone is driving 50 kilometers per hour, that rate can be converted into miles per hour by using a ratio known as the **conversion factor**. Since the given value contains kilometers and the final answer needs to be in miles, the ratio relating miles to kilometers needs to be used. There are 0.62 miles in 1 kilometer. This, written as a ratio and in fraction form, is:

$$\frac{0.62\ miles}{1\ km}$$

To convert 50km/hour into miles per hour, the following conversion needs to be set up:

$$\frac{50\ km}{hour} \times \frac{0.62\ miles}{1\ km} = 31\ miles\ per\ hour$$

The ratio between two similar geometric figures is called the **scale factor**. For example, a problem may depict two similar triangles, A and B. The scale factor from the smaller triangle A to the larger triangle B is given as 2 because the length of the corresponding side of the larger triangle, 16, is twice the corresponding side on the smaller triangle, 8. This scale factor can also be used to find the value of a missing side, x, in triangle A. Since the scale factor from the smaller triangle (A) to larger one (B) is 2, the larger corresponding side in triangle B (given as 25), can be divided by 2 to find the missing side in A ($x = 12.5$). The scale factor can also be represented in the equation $2A = B$ because two times the lengths of A gives the corresponding lengths of B. This is the idea behind similar triangles.

Unit Rate

Unit rate word problems will ask to calculate the rate or quantity of something in a different value. For example, a problem might say that a car drove a certain number of miles in a certain number of minutes and then ask how many miles per hour the car was traveling. These questions involve solving proportions. Consider the following examples:

1) Alexandra made $96 during the first 3 hours of her shift as a temporary worker at a law office. She will continue to earn money at this rate until she finishes in 5 more hours. How much does Alexandra make per hour? How much will Alexandra have made at the end of the day?

This problem can be solved in two ways. The first is to set up a proportion, as the rate of pay is constant. The second is to determine her hourly rate, multiply the 5 hours by that rate, and then add the $96.

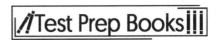

To set up a proportion, put the money already earned over the hours already worked on one side of an equation. The other side has x over 8 hours (the total hours worked in the day). It looks like this:

$$\frac{96}{3} = \frac{x}{8}$$

Now, cross-multiply to get $768 = 3x$. To get x, divide by 3, which leaves $x = 256$. Alternatively, as x is the numerator of one of the proportions, multiplying by its denominator will reduce the solution by one step. Thus, Alexandra will make \$256 at the end of the day. To calculate her hourly rate, divide the total by 8, giving \$32 per hour.

Alternatively, it is possible to figure out the hourly rate by dividing \$96 by 3 hours to get \$32 per hour. Now her total pay can be figured by multiplying \$32 per hour by 8 hours, which comes out to \$256.

2) Jonathan is reading a novel. So far, he has read 215 of the 335 total pages. It takes Jonathan 25 minutes to read 10 pages, and the rate is constant. How long does it take Jonathan to read one page? How much longer will it take him to finish the novel? Express the answer in time.

To calculate how long it takes Jonathan to read one page, divide the 25 minutes by 10 pages to determine the page per minute rate. Thus, it takes 2.5 minutes to read one page.

Jonathan must read 120 more pages to complete the novel. (This is calculated by subtracting the pages already read from the total.) Now, multiply his rate per page by the number of pages. Thus:

$$120 \times 2.5 = 300$$

Expressed in time, 300 minutes is equal to 5 hours.

3) At a hotel, $\frac{4}{5}$ of the 120 rooms are booked for Saturday. On Sunday, $\frac{3}{4}$ of the rooms are booked. On which day are more of the rooms booked, and by how many more?

The first step is to calculate the number of rooms booked for each day. Do this by multiplying the fraction of the rooms booked by the total number of rooms.

Saturday: $\frac{4}{5} \times 120 = \frac{4}{5} \times \frac{120}{1} = \frac{480}{5} = 96$ rooms

Sunday: $\frac{3}{4} \times 120 = \frac{3}{4} \times \frac{120}{1} = \frac{360}{4} = 90$ rooms

Thus, more rooms were booked on Saturday by 6 rooms.

4) In a veterinary hospital, the veterinarian-to-pet ratio is 1:9. The ratio is always constant. If there are 45 pets in the hospital, how many veterinarians are currently in the veterinary hospital?

Set up a proportion to solve for the number of veterinarians: $\frac{1}{9} = \frac{x}{45}$

Cross-multiplying results in $9x = 45$, which works out to 5 veterinarians.

Alternatively, as there are always 9 times as many pets as veterinarians, it is possible to divide the number of pets (45) by 9. This also arrives at the correct answer of 5 veterinarians.

5) At a general practice law firm, 30% of the lawyers work solely on tort cases. If 9 lawyers work solely on tort cases, how many lawyers work at the firm?

First, solve for the total number of lawyers working at the firm, which will be represented here with x. The problem states that 9 lawyers work solely on torts cases, and they make up 30% of the total lawyers at the firm. Thus, 30% multiplied by the total, x, will equal 9. Written as equation, this is:

$$30\% \times x = 9$$

It's easier to deal with the equation after converting the percentage to a decimal, leaving $0.3x = 9$. Thus, $x = \frac{9}{0.3} = 30$ lawyers working at the firm.

6) Xavier was hospitalized with pneumonia. He was originally given 35mg of antibiotics. Later, after his condition continued to worsen, Xavier's dosage was increased to 60mg. What was the percent increase of the antibiotics? Round the percentage to the nearest tenth.

An increase or decrease in percentage can be calculated by dividing the difference in amounts by the original amount and multiplying by 100. Written as an equation, the formula is:

$$\frac{new\ quantity - old\ quantity}{old\ quantity} \times 100$$

Here, the question states that the dosage was increased from 35mg to 60mg, so these are plugged into the formula to find the percentage increase.

$$\frac{60 - 35}{35} \times 100 = \frac{25}{35} \times 100$$

$$0.7142 \times 100 = 71.4\%$$

Analyzing Proportional Relationships and Using Them to Solve Real-World Problems

Much like a scale factor can be written using an equation like $2A = B$, a **relationship** is represented by the equation $Y = kX$. X and Y are proportional because as values of X increase, the values of Y also increase. A relationship that is inversely proportional can be represented by the equation $Y = \frac{k}{X}$, where the value of Y decreases as the value of x increases and vice versa.

Proportional reasoning can be used to solve problems involving ratios, percentages, and averages. Ratios can be used in setting up proportions and solving them to find unknowns. For example, if a student completes an average of 10 pages of math homework in 3 nights, how long would it take the student to complete 22 pages? Both ratios can be written as fractions. The second ratio would contain the unknown.

The following proportion represents this problem, where x is the unknown number of nights:

$$\frac{10\ pages}{3\ nights} = \frac{22\ pages}{x\ nights}$$

Solving this proportion entails cross-multiplying and results in the following equation:

$$10x = 22 \times 3$$

Simplifying and solving for x results in the exact solution: $x = 6.6 \; nights$. The result would be rounded up to 7 because the homework would actually be completed on the 7th night.

The following problem uses ratios involving percentages:

If 20% of the class is girls and 30 students are in the class, how many girls are in the class?

To set up this problem, it is helpful to use the common proportion:

$$\frac{\%}{100} = \frac{is}{of}$$

Within the proportion, % is the percentage of girls, 100 is the total percentage of the class, *is* is the number of girls, and *of* is the total number of students in the class. Most percentage problems can be written using this language. To solve this problem, the proportion should be set up as $\frac{20}{100} = \frac{x}{30}$, and then solved for x. Cross-multiplying results in the equation $20 \times 30 = 100x$, which results in the solution $x = 6$. There are 6 girls in the class.

Problems involving volume, length, and other units can also be solved using ratios. For example, a problem may ask for the volume of a cone to be found that has a radius, $r = 7m$ and a height, $h = 16m$. Referring to the formulas provided on the test, the volume of a cone is given as: $V = \pi r^2 \frac{h}{3}$, where r is the radius, and h is the height. Plugging $r = 7$ and $h = 16$ into the formula, the following is obtained:

$$V = \pi(7^2)\frac{16}{3}$$

Therefore, volume of the cone is found to be approximately 821m³. Sometimes, answers in different units are sought. If this problem wanted the answer in liters, 821m³ would need to be converted. Using the equivalence statement 1m³ = 1000L, the following ratio would be used to solve for liters:

$$821\mathrm{m}^3 \times \frac{1000L}{1m^3}$$

Cubic meters in the numerator and denominator cancel each other out, and the answer is converted to 821,000 liters, or 8.21×10^5 L.

Other conversions can also be made between different given and final units. If the temperature in a pool is 30°C, what is the temperature of the pool in degrees Fahrenheit? To convert these units, an equation is used relating Celsius to Fahrenheit. The following equation is used:

$$T_{°F} = 1.8T_{°C} + 32$$

Plugging in the given temperature and solving the equation for T yields the result:

$$T_{°F} = 1.8(30) + 32 = 86°F$$

Both units in the metric system and U.S. customary system are widely used.

Here are some more examples of how to solve for proportions:

1) $\frac{75\%}{90\%} = \frac{25\%}{x}$

To solve for x, the fractions must be cross multiplied:

$$(75\%x = 90\% \times 25\%)$$

To make things easier, let's convert the percentages to decimals:

$$(0.9 \times 0.25 = 0.225 = 0.75x)$$

To get rid of x's coefficient, each side must be divided by that same coefficient to get the answer:

$$x = 0.3$$

The question could ask for the answer as a percentage or fraction in lowest terms, which are 30% and $\frac{3}{10}$, respectively.

2) $\frac{x}{12} = \frac{30}{96}$

Cross-multiply: $96x = 30 \times 12$

Multiply: $96x = 360$

Divide: $x = 360 \div 96$

Answer: $x = 3.75$

3) $\frac{0.5}{3} = \frac{x}{6}$

Cross-multiply: $3x = 0.5 \times 6$

Multiply: $3x = 3$

Divide: $x = 3 \div 3$

Answer: $x = 1$

You may have noticed there's a faster way to arrive at the answer. If there is an obvious operation being performed on the proportion, the same operation can be used on the other side of the proportion to solve for x. For example, in the first practice problem, 75% became 25% when divided by 3, and upon doing the same to 90%, the correct answer of 30% would have been found with much less legwork. However, these questions aren't always so intuitive, so it's a good idea to work through the steps, even if the answer seems apparent from the outset.

Exponents

Radicals and Exponents

Exponents are used in mathematics to express a number or variable multiplied by itself a certain number of times. For example, x^3 means x is multiplied by itself three times. In this expression, x is called the **base**, and 3 is the **exponent**. Exponents can be used in more complex problems when they contain fractions and negative numbers.

Fractional exponents can be explained by looking first at the inverse of exponents, which are **roots**. Given the expression x^2, the square root can be taken, $\sqrt{x^2}$, cancelling out the 2 and leaving x by itself, if x is positive. Cancellation occurs because $\sqrt{x}$ can be written with exponents, instead of roots, as $x^{\frac{1}{2}}$. The numerator of 1 is the exponent, and the denominator of 2 is called the root (which is why it's referred to as **square root**). Taking the square root of x^2 is the same as raising it to the $\frac{1}{2}$ power. Written out in mathematical form, it takes the following progression:

$$\sqrt{x^2} = (x^2)^{\frac{1}{2}} = x$$

From properties of exponents, $2 \times \frac{1}{2} = 1$ is the actual exponent of x. Another example can be seen with $x^{\frac{4}{7}}$. The variable x, raised to four-sevenths, is equal to the seventh root of x to the fourth power: $\sqrt[7]{x^4}$. In general:

$$x^{\frac{1}{n}} = \sqrt[n]{x}$$

and

$$x^{\frac{m}{n}} = \sqrt[n]{x^m}$$

Negative exponents also involve fractions. Whereas y^3 can also be rewritten as $\frac{y^3}{1}$, y^{-3} can be rewritten as $\frac{1}{y^3}$. A negative exponent means the exponential expression must be moved to the opposite spot in a fraction to make the exponent positive. If the negative appears in the numerator, it moves to the denominator. If the negative appears in the denominator, it is moved to the numerator. In general, $a^{-n} = \frac{1}{a^n}$, and a^{-n} and a^n are reciprocals.

Take, for example, the following expression:

$$\frac{a^{-4}b^2}{c^{-5}}$$

Since a is raised to the negative fourth power, it can be moved to the denominator. Since c is raised to the negative fifth power, it can be moved to the numerator. The b variable is raised to the positive second power, so it does not move.

The simplified expression is as follows:

$$\frac{b^2 c^5}{a^4}$$

In mathematical expressions containing exponents and other operations, the order of operations must be followed. PEMDAS states that exponents are calculated after any parenthesis and grouping symbols, but before any multiplication, division, addition, and subtraction.

There are a few rules for working with exponents. For any numbers a, b, m, n, the following hold true:

$$a^1 = a$$

$$1^a = 1$$

$$a^0 = 1$$

$$a^m \times a^n = a^{m+n}$$

$$a^m \div a^n = a^{m-n}$$

$$(a^m)^n = a^{m \times n}$$

$$(a \times b)^m = a^m \times b^m$$

$$(a \div b)^m = a^m \div b^m$$

Any number, including a fraction, can be an exponent. The same rules apply.

Manipulating Roots and Exponents

A **root** is a different way to write an exponent when the exponent is the reciprocal of a whole number. We use the **radical** symbol to write this in the following way:

$$\sqrt[n]{a} = a^{\frac{1}{n}}$$

This quantity is called the *n-th root* of *a*. The *n* is called the **index** of the radical.

Note that if the *n*-th root of *a* is multiplied by itself *n* times, the result will just be *a*. If no number *n* is written by the radical, it is assumed that *n* is 2:

$$\sqrt{5} = 5^{\frac{1}{2}}$$

The special case of the 2nd root is called the **square root**, and the third root is called the **cube root**.

A **perfect square** is a whole number that is the square of another whole number. For example, sixteen and 64 are perfect squares because 16 is the square of 4, and 64 is the square of 8.

Scientific Notation

Scientific notation is the conversion of extremely small or large numbers into a format that is easier to comprehend and manipulate. It changes the number into three separate parts: a mathematical sign ($+/-$), a digit term (known as a **significand**), and an exponential term.

Scientific notation = (+ or -) significand x exponential term

To put a number into scientific notation, one should use the following steps:

- Move the decimal point to after the first non-zero number to find the digit number.
- Count how many places the decimal point was moved in step 1.
- Determine if the exponent is positive or negative.
- Create an exponential term using the information from steps 2 and 3.
- Combine the digit term and exponential term to get scientific notation.

For example, to put 0.0000098 into scientific notation, the decimal should be moved so that it lies between the last two numbers: 000009.8. This creates the digit number:

$$9.8$$

Next, the number of places that the decimal point moved is determined; to get between the 9 and the 8, the decimal was moved six places to the right. It may be helpful to remember that a decimal moved to the right creates a negative exponent, and a decimal moved to the left creates a positive exponent. Because the decimal was moved six places to the right, the exponent is negative.

Now, the exponential term can be created by using the base 10 (this is *always* the base in scientific notation) and the number of places moved as the exponent, in this case:

$$10^{-6}$$

Finally, the digit term and the exponential term can be combined as a product. Therefore, the scientific notation for the number 0.0000098 is:

$$9.8 \times 10^{-6}$$

Algebraic Expressions

Algebraic Expressions

Algebraic expressions look similar to equations, but they do not include the equal sign. Algebraic expressions are comprised of numbers, variables, and mathematical operations. Some examples of algebraic expressions are:

$$8x + 7y - 12z, 3a^2$$

and

$$5x^3 - 4y^4.$$

Algebraic expressions consist of variables, numbers, and operations. A **term** of an expression is any combination of numbers and/or variables, and terms are separated by addition and subtraction. For example, the expression:

$$5x^2 - 3xy + 4 - 2$$

consists of 4 terms: $5x^2$, -3xy, 4y, and -2. Note that each term includes its given sign (+ or −). The **variable** part of a term is a letter that represents an unknown quantity. The coefficient of a term is the number by which the variable is multiplied. For the term 4y, the variable is y and the coefficient is 4. Terms are identified by the power (or exponent) of its variable.

A number without a variable is referred to as a **constant**. If the variable is to the first power (x^1 or simply x), it is referred to as a **linear** term. A term with a variable to the second power (x^2) is **quadratic** and a term to the third power (x^3) is **cubic**. Consider the expression:

$$x^3 + 3x - 1$$

The constant is -1. The linear term is 3x. There is no quadratic term. The cubic term is x^3.

An algebraic expression can also be classified by how many terms exist in the expression. Any like terms should be combined before classifying. A **monomial** is an expression consisting of only one term. Examples of monomials are: 17, 2x, and $-5ab^2$. A **binomial** is an expression consisting of two terms separated by addition or subtraction. Examples include $2x - 4$ and $-3y^2 + 2y$. A **trinomial** consists of 3 terms. For example, $5x^2 - 2x + 1$ is a trinomial.

Algebraic expressions and equations can be used to represent real-life situations and model the behavior of different variables. For example, $2x + 5$ could represent the cost to play games at an arcade. In this case, 5 represents the price of admission to the arcade and 2 represents the cost of each game played. To calculate the total cost, use the number of games played for x, multiply it by 2, and add 5.

Adding and Subtracting Linear Algebraic Expressions

An algebraic expression is simplified by combining like terms. A term is a number, variable, or product of a number, and variables separated by addition and subtraction. For the algebraic expression:

$$3x^2 - 4x + 5 - 5x^2 + x - 3$$

the terms are $3x^2$, -4x, 5, -5x^2, x, and -3. **Like terms** have the same variables raised to the same powers (exponents). The like terms for the previous example are $3x^2$ and -5x^2, -4x and x, 5 and -3.

To combine like terms, the coefficients (numerical factor of the term including sign) are added, and the variables and their powers are kept the same. Note that if a coefficient is not written, it is an implied coefficient of 1 ($x = 1x$). The previous example will simplify to $-2x^2 - 3x + 2$.

When adding or subtracting algebraic expressions, each expression is written in parenthesis. The negative sign is distributed when necessary, and like terms are combined. Consider the following:

$$\text{add } 2a + 5b - 2 \text{ to } a - 2b + 8c - 4$$

The sum is set as follows:

$$(a - 2b + 8c - 4) + (2a + 5b - 2)$$

In front of each set of parentheses is an implied positive one, which, when distributed, does not change any of the terms. Therefore, the parentheses are dropped and like terms are combined:

$$a - 2b + 8c - 4 + 2a + 5b - 2$$

$$3a + 3b + 8c - 6$$

Consider the following problem:

$$\text{Subtract } 2a + 5b - 2 \text{ from } a - 2b + 8c - 4$$

The difference is set as follows:

$$(a - 2b + 8c - 4) - (2a + 5b - 2)$$

The implied one in front of the first set of parentheses will not change those four terms. However, distributing the implied -1 in front of the second set of parentheses will change the sign of each of those three terms:

$$a - 2b + 8c - 4 - 2a - 5b + 2$$

Combining like terms yields the simplified expression:

$$-a - 7b + 8c - 2$$

Multiplying Algebraic Expressions

The **distributive property** states that multiplying a sum (or difference) by a number produces the same result as multiplying each value in the sum (or difference) by the number and adding (or subtracting) the products. Using mathematical symbols, the distributive property states $a(b + c) = ab + ac$. The expression $4(3 + 2)$ is simplified using the order of operations. Simplifying inside the parenthesis first produces 4×5, which equals 20. The expression $4(3 + 2)$ can also be simplified using the distributive property:

$$4(3 + 2)$$

$$4 \times 3 + 4 \times 2$$

$$12 + 8$$

$$20$$

Consider the following example: $4(3x - 2)$. The expression cannot be simplified inside the parenthesis because $3x$ and -2 are not like terms, and therefore cannot be combined. However, the expression can be simplified by using the distributive property and multiplying each term inside of the parenthesis by the term outside of the parenthesis: $12x - 8$. The resulting equivalent expression contains no like terms, so it cannot be further simplified.

Consider the expression:

$$(3x + 2y + 1) - (5x - 3) + 2(3y + 4)$$

Again, there are no like terms, but the distributive property is used to simplify the expression. Note there is an implied one in front of the first set of parentheses and an implied -1 in front of the second set of parentheses. Distributing the one, -1, and 2 produces:

$$1(3x) + 1(2y) + 1(1) - 1(5x) - 1(-3) + 2(3y) + 2(4)$$

$$3x + 2y + 1 - 5x + 3 + 6y + 8$$

This expression contains like terms that are combined to produce the simplified expression:

$$-2x + 8y + 12$$

Algebraic expressions are tested to be equivalent by choosing values for the variables and evaluating both expressions. For example, $4(3x - 2)$ and $12x - 8$ are tested by substituting 3 for the variable x and calculating to determine if equivalent values result.

Evaluating Algebraic Expressions

To evaluate the expression, the given values for the variables are substituted (or replaced) and the expression is simplified using the order of operations. Parenthesis should be used when substituting. Consider the following: Evaluate $a - 2b + ab$ for $a = 3$ and $b = -1$. To evaluate, any variable a is replaced with 3 and any variable b with -1, producing:

$$(3) - 2(-1) + (3)(-1)$$

Next, the order of operations is used to calculate the value of the expression, which is 2.

Here's another example:

$$\text{Evaluate } a - 2b + ab \ for \ a = 3 \text{ and } b = -1$$

To evaluate an expression, the given values should be substituted for the variables and simplified using the order of operations. In this case:

$$(3) - 2(-1) + (3)(-1)$$

Parentheses are used when substituting.

Given an algebraic expression, students may be asked to simplify the expression. For example:

$$\text{Simplify } 5x^2 - 10x + 2 - 8x^2 + x - 1.$$

Simplifying algebraic expressions requires combining like terms. A term is a number, variable, or product of a number and variables separated by addition and subtraction. The terms in the above expression are: $5x^2, -10x, 2, -8x^2, x,$ and -1. Like terms have the same variables raised to the same powers (exponents). To combine like terms, the coefficients (numerical factor of the term including sign) are added, while the variables and their powers are kept the same. The example above simplifies to:

$$-3x^2 - 9x + 1$$

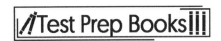

Let's try two more.

Evaluate $\frac{1}{2}x^2 - 3, x = 4$.

The first step is to substitute in 4 for x in the expression:

$$\frac{1}{2}(4)^2 - 3$$

Then, the order of operations is used to simplify.

The exponent comes first, $\frac{1}{2}(16) - 3$, then the multiplication $8 - 3$, and then, after subtraction, the solution is 5.

Evaluate $4|5 - x| + 2y, x = 4, y = -3$.

The first step is to substitute 4 in for x and -3 in for y in the expression:

$$4|5 - 4| + 2(-3)$$

Then, the absolute value expression is simplified, which is:

$$|5 - 4| = |1| = 1$$

The expression is $4(1) + 2(-3)$ which can be simplified using the order of operations.

First is the multiplication, $4 + (-6)$; then addition yields an answer of -2.

Creating Algebraic Expressions

A linear expression is a statement about an unknown quantity expressed in mathematical symbols. The statement "five times a number added to forty" can be expressed as $5x + 40$. A linear equation is a statement in which two expressions (at least one containing a variable) are equal to each other. The statement "five times a number added to forty is equal to ten" can be expressed as $5x + 40 = 10$.

Real world scenarios can also be expressed mathematically. Suppose a job pays its employees $300 per week and $40 for each sale made. The weekly pay is represented by the expression $40x + 300$ where x is the number of sales made during the week.

Consider the following scenario: Bob had $20 and Tom had $4. After selling 4 ice cream cones to Bob, Tom has as much money as Bob. The cost of an ice cream cone is an unknown quantity and can be represented by a variable (x). The amount of money Bob has after his purchase is four times the cost of an ice cream cone subtracted from his original $20 → $20 - 4x$. The amount of money Tom has after his sale is four times the cost of an ice cream cone added to his original $4 → $4x + 4$. After the sale, the amount of money that Bob and Tom have are equal → $20 - 4x = 4x + 4$.

Test Prep Books!!!

When expressing a verbal or written statement mathematically, it is key to understand words or phrases that can be represented with symbols. The following are examples:

Symbol	Phrase
$+$	added to, increased by, sum of, more than
$-$	decreased by, difference between, less than, take away
x	multiplied by, 3 (4, 5 . . .) times as large, product of
$\div$	divided by, quotient of, half (third, etc.) of
$=$	is, the same as, results in, as much as
$x, t, n, etc.$	a number, unknown quantity, value of

Linear Equations

One-Variable Linear Equations and Inequalities

Linear equations and linear inequalities are both comparisons of two algebraic expressions. However, unlike equations in which the expressions are equal, linear inequalities compare expressions that may be unequal. Linear equations typically have one value for the variable that makes the statement true. Linear inequalities generally have an infinite number of values that make the statement true.

When solving a linear equation, the desired result requires determining a numerical value for the unknown variable. If given a linear equation involving addition, subtraction, multiplication, or division, working backwards isolates the variable. Addition and subtraction are inverse operations, as are multiplication and division. Therefore, they can be used to cancel each other out.

For example, solve $4(t - 2) + 2t - 4 = 2(9 - 2t)$

Distributing: $4t - 8 + 2t - 4 = 18 - 4t$

Combining like terms: $6t - 12 = 18 - 4t$

Adding $4t$ to each side to move the variable: $10t - 12 = 18$

Adding 12 to each side to isolate the variable: $10t = 30$

Dividing each side by 10 to isolate the variable: $t = 3$

The answer can be checked by substituting the value for the variable into the original equation, ensuring that both sides calculate to be equal.

Linear inequalities express the relationship between unequal values. More specifically, they describe in what way the values are unequal. A value can be greater than (>), less than (<), greater than or equal to (≥), or less than or equal to (≤) another value. $5x + 40 > 65$ is read as *five times a number added to forty is greater than sixty-five.*

When solving a linear inequality, the solution is the set of all numbers that make the statement true. The inequality $x + 2 \geq 6$ has a solution set of 4 and every number greater than 4 (4.01; 5; 12; 107; etc.). Adding 2 to 4 or any number greater than 4 results in a value that is greater than or equal to 6. Therefore, $x \geq 4$ is the solution set.

54

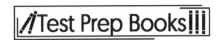

To algebraically solve a linear inequality, follow the same steps as those for solving a linear equation. The inequality symbol stays the same for all operations *except* when multiplying or dividing by a negative number. If multiplying or dividing by a negative number while solving an inequality, the relationship reverses (the sign flips). In other words, > switches to < and vice versa. Multiplying or dividing by a positive number does not change the relationship, so the sign stays the same. An example is shown below.

Solve $-2x - 8 \leq 22$

Add 8 to both sides: $-2x \leq 30$

Divide both sides by -2: $x \geq -15$

Solutions of a linear equation or a linear inequality are the values of the variable that make a statement true. In the case of a linear equation, the solution set (list of all possible solutions) typically consists of a single numerical value. To find the solution, the equation is solved by isolating the variable. For example, solving the equation $3x - 7 = -13$ produces the solution $x = -2$. The only value for x which produces a true statement is -2. This can be checked by substituting -2 into the original equation to check that both sides are equal. In this case,

$$3(-2) - 7 = -13 \rightarrow -13 = -13;$$

therefore, -2 is a solution.

Although linear equations generally have one solution, this is not always the case. If there is no value for the variable that makes the statement true, there is no solution to the equation. Consider the equation $x + 3 = x - 1$. There is no value for x in which adding 3 to the value produces the same result as subtracting one from the value. Conversely, if any value for the variable makes a true statement, the equation has an infinite number of solutions. Consider the equation:

$$3x + 6 = 3(x + 2)$$

Any number substituted for x will result in a true statement (both sides of the equation are equal).

By manipulating equations like the two above, the variable of the equation will cancel out completely. If the remaining constants express a true statement (ex. $6 = 6$), then all real numbers are solutions to the equation. If the constants left express a false statement (ex. $3 = -1$), then no solution exists for the equation.

Solving a linear inequality requires all values that make the statement true to be determined. For example, solving $3x - 7 \geq -13$ produces the solution $x \geq -2$. This means that -2 and any number greater than -2 produces a true statement. Solution sets for linear inequalities will often be displayed using a number line. If a value is included in the set ($\geq$ or $\leq$), a shaded dot is placed on that value and an arrow extending in the direction of the solutions. For a variable > or $\geq$ a number, the arrow will point right on a number line, the direction where the numbers increase. If a variable is < or $\leq$ a number, the arrow will point left on a number line, which is the direction where the numbers decrease. If the value is not included in the set (> or <), an open (unshaded) circle on that value is used with an arrow in the appropriate direction.

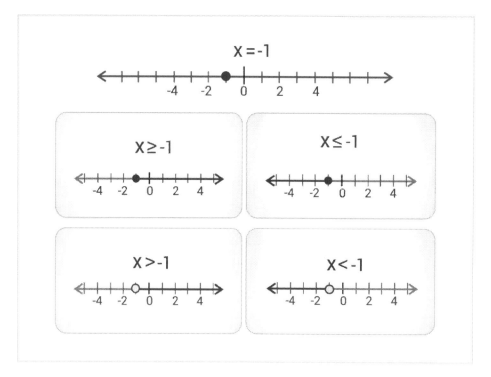

Similar to linear equations, a linear inequality may have a solution set consisting of all real numbers or can contain no solution. When solved algebraically, a linear inequality in which the variable cancels out and results in a true statement (ex. $7 \geq 2$) has a solution set of all real numbers. A linear inequality in which the variable cancels out and results in a false statement (ex. $7 \leq 2$) has no solution.

Linear Inequalities in One Variable

Linear inequalities and linear equations are both comparisons of two algebraic expressions. However, unlike equations in which the expressions are equal to each other, **linear inequalities** compare expressions that are unequal. Linear equations typically have one value for the variable that makes the statement true. Linear inequalities generally have an infinite number of values that make the statement true.

Linear inequalities are a concise mathematical way to express the relationship between unequal values. More specifically, they describe in what way the values are unequal. A value could be greater than (>); less than (<); greater than or equal to ($\geq$); or less than or equal to ($\leq$) another value. The statement "five times a number added to forty is more than sixty-five" can be expressed as $5x + 40 > 65$. Common words and phrases that express inequalities are:

Symbol	Phrase
<	is under, is below, smaller than, beneath
>	is above, is over, bigger than, exceeds
$\leq$	no more than, at most, maximum
$\geq$	no less than, at least, minimum

Solving Linear Inequalities

When solving a linear inequality, the solution is the set of all numbers that makes the statement true. The inequality $x + 2 \geq 6$ has a solution set of 4 and every number greater than 4 (4.0001, 5, 12, 107, etc.). Adding 2 to 4 or any number greater than 4 would result in a value that is greater than or equal to 6. Therefore, $x \geq 4$ would be the solution set.

Solution sets for linear inequalities often will be displayed using a number line. If a value is included in the set ($\geq$ or $\leq$), there is a shaded dot placed on that value and an arrow extending in the direction of the solutions. For a variable $>$ or $\geq$ a number, the arrow would point right on the number line (the direction where the numbers increase); and if a variable is $<$ or $\leq$ a number, the arrow would point left (where the numbers decrease). If the value is not included in the set ($>$ or $<$), an open circle on that value would be used with an arrow in the appropriate direction.

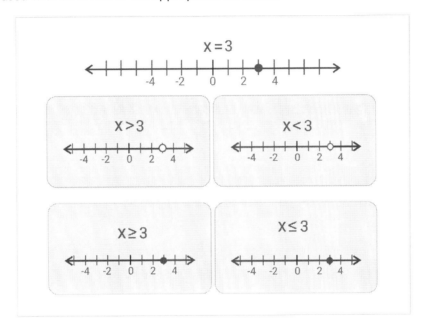

Students may be asked to write a linear inequality given a graph of its solution set. To do so, they should identify whether the value is included (shaded dot or open circle) and the direction in which the arrow is pointing.

In order to algebraically solve a linear inequality, the same steps should be followed as in solving a linear equation. The inequality symbol stays the same for all operations EXCEPT when multiplying or dividing by a negative number. If multiplying or dividing by a negative number while solving an inequality, the relationship reverses (the sign flips). Multiplying or dividing by a positive does not change the relationship, so the sign stays the same. In other words, $>$ switches to $<$ and vice versa. An example is shown below.

Solve $-2(x + 4) \leq 22$

Distribute: $-2x - 8 \leq 22$

Add 8 to both sides: $-2x \leq 30$

Divide both sides by -2: $x \geq 15$

Systems of Two Linear Equations in Two Variables

A system of two linear equations in two variables is a set of equations that use the same variables, usually *x* and *y*. Here's a sample problem:

> An Internet provider charges an installation fee and a monthly charge. It advertises that two months of its offering costs $100 and six months costs $200. Find the monthly charge and the installation fee.

The two unknown quantities (variables) are the monthly charge and the installation fee. There are two different statements given relating the variables: two months added to the installation fee is $100; and six months added to the installation fee is $200. Using the variable *x* as the monthly charge and *y* as the installation fee, the statements can be written as the following: $2x + y = 100$; $6x + y = 200$. These two equations taken together form a system modeling the given scenario.

Systems of Linear Inequalities in Two Variables

A system of linear inequalities consists of two linear inequalities making comparisons between two variables. Students may be given a scenario and asked to express it as a system of inequalities:

> A consumer study calls for at least 60 adult participants. It cannot use more than 25 men. Express these constraints as a system of inequalities.

This can be modeled by the system: $x + y \geq 60$; $x \leq 25$, where *x* represents the number of men and *y* represents the number of women. A solution to the system is an ordered pair that makes both inequalities true when substituting the values for *x* and *y*.

Linear Applications and Graphs

Linear Models and Relationships

Linear relationships describe the way two quantities change with respect to each other. The relationship is defined as linear because a line is produced if all the sets of corresponding values are graphed on a coordinate grid. When expressing the linear relationship as an equation, the equation is often written in the form $y = mx + b$ (slope-intercept form) where *m* and *b* are numerical values and *x* and *y* are variables (for example, $y = 5x + 10$). Given a linear equation and the value of either variable (*x* or *y*), the value of the other variable can be determined.

Suppose a teacher is grading a test containing 20 questions with 5 points given for each correct answer, adding a curve of 10 points to each test. This linear relationship can be expressed as the equation $y = 5x + 10$ where *x* represents the number of correct answers and *y* represents the test score. To determine the score of a test with a given number of correct answers, the number of correct answers is substituted into the equation for *x* and evaluated. For example, for 10 correct answers, 10 is substituted for *x*:

$$y = 5(10) + 10 \rightarrow y = 60$$

Therefore, 10 correct answers will result in a score of 60. The number of correct answers needed to obtain a certain score can also be determined. To determine the number of correct answers needed to score a 90, 90 is substituted for y in the equation (y represents the test score) and solved:

$$90 = 5x + 10 \rightarrow 80 = 5x \rightarrow 16 = x$$

Therefore, 16 correct answers are needed to score a 90.

Linear relationships may be represented by a table of 2 corresponding values. Certain tables may determine the relationship between the values and predict other corresponding sets. Consider the table below, which displays the money in a checking account that charges a monthly fee:

Month	0	1	2	3	4
Balance	$210	$195	$180	$165	$150

An examination of the values reveals that the account loses $15 every month (the month increases by one and the balance decreases by 15). This information can be used to predict future values. To determine what the value will be in month 6, the pattern can be continued, and it can be concluded that the balance will be $120. To determine which month the balance will be $0, $210 is divided by $15 (since the balance decreases $15 every month), resulting in month 14.

Similar to a table, a graph can display corresponding values of a linear relationship.

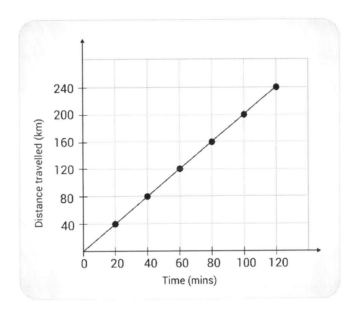

The graph above represents the relationship between distance traveled and time. To find the distance traveled in 80 minutes, the mark for 80 minutes is located at the bottom of the graph. By following this mark directly up on the graph, the corresponding point for 80 minutes is directly across from the 160-kilometer mark. This information indicates that the distance travelled in 80 minutes is 160 kilometers. To predict information not displayed on the graph, the way in which the variables change with respect to one another is determined. In this case, distance increases by 40 kilometers as time increases by 20 minutes. This information can be used to continue the data in the graph or convert the values to a table.

Identifying Variables for Linear Models

The first step to writing a linear model is to identify what the variables represent. A **variable** represents an unknown quantity, and in the case of a linear equation, a specific relationship exists between the two variables (usually *x* and *y*). Within a given scenario, the variables are the two quantities that are changing. The variable *x* is considered the independent variable and represents the inputs of a function. The variable *y* is considered the dependent variable and represents the outputs of a function. For example, if a scenario describes distance traveled and time traveled, distance would be represented by *y* and time represented by *x*. The distance traveled depends on the time spent traveling (time is independent). If a scenario describes the cost of a cab ride and the distance traveled, the cost would be represented by *y* and the distance represented by *x*. The cost of a cab ride depends on the distance traveled.

Identifying the Slope and Y-Intercept for Linear Models

The **slope of the graph of a line** represents the rate of change between the variables of an equation. In the context of a real-world scenario, the slope will tell the way in which the unknown quantities (variables) change with respect to each other. A scenario involving distance and time might state that someone is traveling at a rate of 45 miles per hour. The slope of the linear model would be 45. A scenario involving the cost of a cab ride and distance traveled might state that the person is charged $3 for each mile. The slope of the linear model would be 3.

The **y-intercept of a linear function** is the value of *y* when $x = 0$ (the point where the line intercepts the *y*-axis on the graph of the equation). It is sometimes helpful to think of this as a "starting point" for a linear function. Suppose for the scenario about the cab ride that the person is told that the cab company charges a flat fee of $5 plus $3 for each mile. Before traveling any distance ($x = 0$), the cost is $5. The *y*-intercept for the linear model would be 5.

Identifying Ordered Pairs for Linear Models

A linear equation with two variables can be written given a point (ordered pair) and the slope or given two points on a line. An ordered pair gives a set of corresponding values for the two variables (*x* and *y*). As an example, for a scenario involving distance and time, it is given that the person traveled 112.5 miles in 2 ½ hours. Knowing that *x* represents time and *y* represents distance, this information can be written as the ordered pair (2.5, 112.5).

Linear Functions Model a Linear Relationship Between Two Quantities

Linear relationships between two quantities can be expressed in two ways: function notation or as a linear equation with two variables. The relationship is referred to as linear because its graph is represented by a line. For a relationship to be linear, both variables must be raised to the first power only.

Function/Linear Equation Notation

A **relation** is a set of input and output values that can be written as ordered pairs. A function is a relation in which each input is paired with exactly one output. The domain of a function consists of all inputs, and the range consists of all outputs. Graphing the ordered pairs of a linear function produces a straight line. An example of a function would be $f(x) = 4x + 4$, read "*f* of *x* is equal to four times *x* plus four." In this example, the input would be *x* and the output would be f(x). Ordered pairs would be represented as (*x*, f(x)). To find the output for an input value of 3, 3 would be substituted for *x* into the function as follows: $f(3) = 4(3) + 4$, resulting in $f(3) = 16$. Therefore, the ordered pair $(3, f(3)) = (3, 16)$. Note

$f(x)$ is a function of x denoted by f. Functions of x could be named $g(x)$, read "g of x"; $p(x)$, read "p of x"; etc.

A linear function could also be written in the form of an equation with two variables. Typically, the variable x represents the inputs and the variable y represents the outputs. The variable x is considered the independent variable and y the dependent variable. The above function would be written as $y = 4x + 4$. Ordered pairs are written in the form (x, y).

Writing Linear Equations in Two Variables

When writing linear equations in two variables, the process depends on the information given. Questions will typically provide the slope of the line and its y-intercept, an ordered pair and the slope, or two ordered pairs.

Linear equations are commonly written in slope-intercept form, $y = mx + b$, where m represents the slope of the line and b represents the y-intercept. The slope is the rate of change between the variables, usually expressed as a whole number or fraction. The y-intercept is the value of y when $x = 0$ (the point where the line intercepts the y-axis on a graph). Given the slope and y-intercept of a line, the values are substituted for m and b into the equation. A line with a slope of $\frac{1}{2}$ and y-intercept of -2 would have an equation:

$$y = \frac{1}{2}x - 2$$

The point-slope form of a line, $y - y_1 = m(x - x_1)$, is used to write an equation when given an ordered pair (point on the equation's graph) for the function and its rate of change (slope of the line). The values for the slope, m, and the point (x_1, y_1) are substituted into the point-slope form to obtain the equation of the line. A line with a slope of 3 and an ordered pair (4, -2) would have an equation:

$$y - (-2) = 3(x - 4)$$

If a question specifies that the equation be written in slope-intercept form, the equation should be manipulated to isolate y:

Solve: $y - (-2) = 3(x - 4)$

Distribute: $y + 2 = 3x - 12$

Subtract 2 from both sides: $y = 3x - 14$

Given two ordered pairs for a function, (x_1, y_1) and (x_2, y_2), it is possible to determine the rate of change between the variables (slope of the line). To calculate the slope of the line, m, the values for the ordered pairs should be substituted into the formula:

$$m = \frac{y_2 - y_1}{x_2 - x_1}$$

The expression is substituted to obtain a whole number or fraction for the slope. Once the slope is calculated, the slope and either of the ordered pairs should be substituted into the point-slope form to obtain the equation of the line.

Graphing a System of Two Linear Equations in Two Variables

A solution for a system of equations is an ordered pair that makes both equations true. One method for solving a system of equations is to graph both lines on a coordinate plane. If the lines intersect, the point of intersection is the solution to the system. Every point on a line represents an ordered pair that makes its equation true. The ordered pair represented by this point of intersection lies on both lines and therefore makes both equations true. This ordered pair should be checked by substituting its values into both of the original equations of the system. Note that given a system of equations and an ordered pair, the ordered pair can be determined to be a solution or not by checking it in both equations.

If, when graphed, the lines representing the equations of a system do not intersect, then the two lines are parallel to each other or they are the same exact line. Parallel lines extend in the same direction without ever meeting. A system consisting of parallel lines has no solution. If the equations for a system represent the same exact line, then every point on the line is a solution to the system. In this case, there would be an infinite number of solutions. A system consisting of intersecting lines is referred to as independent; a system consisting of parallel lines is referred to as inconsistent; and a system consisting of coinciding lines is referred to as dependent.

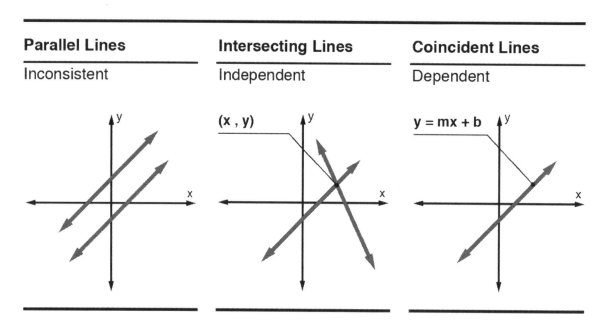

Graphing Solution Sets for Linear Inequalities in Two Variables

A graph of the solution set for a linear inequality shows the ordered pairs that make the statement true. The graph consists of a boundary line dividing the coordinate plane and shading on one side of the boundary. The boundary line should be graphed just as a linear equation would be graphed. If the inequality symbol is $>$ or $<$, a dashed line can be used to indicate that the line is not part of the solution set. If the inequality symbol is $\geq$ or $\leq$, a solid line can be used to indicate that the boundary line is included in the solution set. An ordered pair (x, y) on either side of the line should be chosen to test in the inequality statement. If substituting the values for x and y results in a true statement $(15(3) + 25(2) > 90)$, that ordered pair and all others on that side of the boundary line are part of the solution set. To indicate this, that region of the graph should be shaded. If substituting the ordered pair results in a false statement, the ordered pair and all others on that side are not part of the solution set.

62

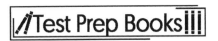

Therefore, the other region of the graph contains the solutions and should be shaded.

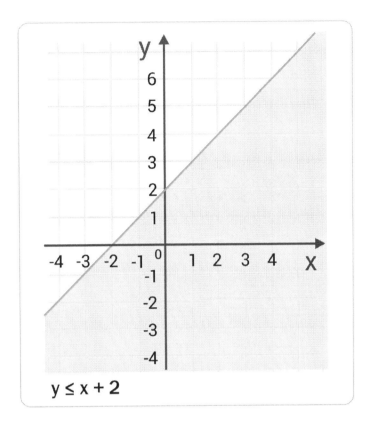

$$y \leq x + 2$$

A question may simply ask whether a given ordered pair is a solution to a given inequality. To determine this, the values should be substituted for the ordered pair into the inequality. If the result is a true statement, the ordered pair is a solution; if the result is a false statement, the ordered pair is not a solution.

Probability Sets

Chance Processes and Probability Models

Counting Techniques

There are many counting techniques that can help solve problems involving counting possibilities. For example, the **Addition Principle** states that if there are m choices from Group 1 and n choices from Group 2, then $n + m$ is the total number of choices possible from Groups 1 and 2. For this to be true, the groups can't have any choices in common. The **Multiplication Principle** states that if Process 1 can be completed n ways and Process 2 can be completed m ways, the total number of ways to complete both Process 1 and Process 2 is $n \times m$. For this rule to be used, both processes must be independent of each other. Counting techniques also involve permutations. A **permutation** is an arrangement of elements in a set for which order must be considered. For example, if three letters from the alphabet are chosen, ABC and BAC are two different permutations. The multiplication rule can be used to determine the total number of possibilities. If each letter can't be selected twice, the total number of possibilities is:

$$26 \times 25 \times 24 = 15{,}600$$

A formula can also be used to calculate this total. In general, the notation $P(n, r)$ represents the number of ways to arrange r objects from a set of n and, the formula is:

$$P(n, r) = \frac{n!}{(n-r)!}$$

In the previous example:

$$P(26, 3) = \frac{26!}{23!} = 15,600$$

Contrasting permutations, a **combination** is an arrangement of elements in which order doesn't matter. In this case, ABC and BAC are the same combination. In the previous scenario, there are six permutations that represent each single combination. Therefore, the total number of possible combinations is:

$$15,600 \div 6 = 2,600$$

In general, $C(n, r)$ represents the total number of combinations of n items selected r at a time where order doesn't matter, and the formula is:

$$C(n, r) = \frac{n!}{(n-r)! \; r!}$$

Therefore, the following relationship exists between permutations and combinations:

$$C(n, r) = \frac{P(n, r)}{n!} = \frac{P(n, r)}{P(r, r)}$$

Fundamental Counting Principle

The **fundamental counting principle** states that if there are m possible ways for an event to occur, and n possible ways for a second event to occur, there are $m \cdot n$ possible ways for both events to occur. For example, there are two events that can occur after flipping a coin and six events that can occur after rolling a die, so there are $2 \times 6 = 12$ total possible event scenarios if both are done simultaneously. This principle can be used to find probabilities involving finite sample spaces and independent trials because it calculates the total number of possible outcomes. For this principle to work, the events must be independent of each other.

Independence and Conditional Probability

Sample Subsets

A sample can be broken up into subsets that are smaller parts of the whole. For example, consider a sample population of females. The sample can be divided into smaller subsets based on the characteristics of each female. There can be a group of females with brown hair and a group of females that wear glasses. There also can be a group of females that have brown hair *and* wear glasses. This "and" relates to the **intersection** of the two separate groups of brunettes and those with glasses. Every female in that intersection group has both characteristics. Similarly, there also can be a group of females that either have brown hair *or* wear glasses. The "or" relates to the union of the two separate groups of brunettes and glasses. Every female in this group has at least one of the characteristics. Finally, the group of females who do not wear glasses can be discussed. This "not" relates to the **complement** of the glass-wearing group. No one in the complement has glasses. **Venn diagrams** are useful in highlighting

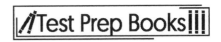

these ideas. When discussing statistical experiments, this idea can also relate to events instead of characteristics.

Verifying Independent Events
Two events aren't always independent. For example, females with glasses and brown hair aren't independent characteristics. There definitely can be overlap because females with brown hair can wear glasses. Also, two events that exist at the same time don't have to have a relationship. For example, even if all females in a given sample are wearing glasses, the characteristics aren't related. In this case, the probability of a brunette wearing glasses is equal to the probability of a female being a brunette multiplied by the probability of a female wearing glasses. This mathematical test of $P(A \cap B) = P(A)P(B)$ verifies that two events are independent.

Conditional Probability
Conditional probability is the probability that event A will happen given that event B has already occurred. An example of this is calculating the probability that a person will eat dessert once they have eaten dinner. This is different than calculating the probability of a person just eating dessert. The formula for the conditional probability of event A occurring given B is:

$$P(A|B) = \frac{P(A \text{ and } B)}{P(B)}$$

and it's defined to be the probability of both A and B occurring divided by the probability of event B occurring. If A and B are independent, then the probability of both A and B occurring is equal to $P(A)P(B)$, so $P(A|B)$ reduces to just $P(A)$. This means that A and B have no relationship, and the probability of A occurring is the same as the conditional probability of A occurring given B. Similarly:

$$P(B|A) = \frac{P(B \text{ and } A)}{P(A)} = P(B)$$

if A and B are independent.

Independent Versus Related Events
To summarize, conditional probability is the probability that an event occurs given that another event has happened. If the two events are related, the probability that the second event will occur changes if the other event has happened. However, if the two events aren't related and are therefore independent, the first event to occur won't impact the probability of the second event occurring.

Measuring Probabilities with Two-Way Frequency Tables
When measuring event probabilities, two-way frequency tables can be used to report the raw data and then used to calculate probabilities. If the frequency tables are translated into relative frequency tables, the probabilities presented in the table can be plugged directly into the formulas for conditional probabilities. By plugging in the correct frequencies, the data from the table can be used to determine if events are independent or dependent.

Differing Probabilities
The probability that event A occurs differs from the probability that event A occurs given B. When working within a given model, it's important to note the difference. $P(A|B)$ is determined using the formula $P(A|B) = \frac{P(A \text{ and } B)}{P(B)}$ and represents the total number of A's outcomes left that could occur

after B occurs. $P(A)$ can be calculated without any regard for B. For example, the probability of a student finding a parking spot on a busy campus is different once class is in session.

The Addition Rule

The probability of event A or B occurring isn't equal to the sum of each individual probability. The probability that both events can occur at the same time must be subtracted from this total. This idea is shown in the **addition rule**:

$$P(A \text{ or } B) = P(A) + P(B) - P(A \text{ and } B)$$

The addition rule is another way to determine the probability of compound events that aren't mutually exclusive. If the events are mutually exclusive, the probability of both A and B occurring at the same time is 0.

Computing Probabilities

Simple and Compound Events

A **simple event** consists of only one outcome. The most popular simple event is flipping a coin, which results in either heads or tails. A **compound event** results in more than one outcome and consists of more than one simple event. An example of a compound event is flipping a coin while tossing a die. The result is either heads or tails on the coin and a number from one to six on the die. The probability of a simple event is calculated by dividing the number of possible outcomes by the total number of outcomes. Therefore, the probability of obtaining heads on a coin is $\frac{1}{2}$, and the probability of rolling a 6 on a die is $\frac{1}{6}$. The probability of compound events is calculated using the basic idea of the probability of simple events. If the two events are independent, the probability of one outcome is equal to the product of the probabilities of each simple event. For example, the probability of obtaining heads on a coin and rolling a 6 is equal to:

$$\frac{1}{2} \times \frac{1}{6} = \frac{1}{12}$$

The probability of either A or B occurring is equal to the sum of the probabilities minus the probability that both A and B will occur. Therefore, the probability of obtaining either heads on a coin or rolling a 6 on a die is:

$$\frac{1}{2} + \frac{1}{6} - \frac{1}{12} = \frac{7}{12}$$

The two events aren't mutually exclusive because they can happen at the same time. If two events are mutually exclusive, and the probability of both events occurring at the same time is zero, the probability of event A or B occurring equals the sum of both probabilities. An example of calculating the probability of two mutually exclusive events is determining the probability of pulling a king or a queen from a deck of cards. The two events cannot occur at the same time.

Uniform and Non-Uniform Probability Models

A **uniform probability model** is one where each outcome has an equal chance of occurring, such as the probabilities of rolling each side of a die. A **non-uniform probability model** is one where each outcome has an unequal chance of occurring. In a uniform probability model, the conditional probability formulas for $P(B|A)$ and $P(A|B)$ can be multiplied by their respective denominators to obtain two formulas for $P(A \text{ and } B)$. Therefore, the multiplication rule is derived as:

$$P(A \text{ and } B) = P(A)P(B|A) = P(B)P(A|B)$$

In a model, if the probability of either individual event is known and the corresponding conditional probability is known, the multiplication rule allows the probability of the joint occurrence of A and B to be calculated.

Descriptive Statistics

Basic Statistics

The field of statistics describes relationships between quantities that are related, but not necessarily in a deterministic manner. For example, a graduating student's salary will often be higher when the student graduates with a higher GPA, but this is not always the case. Likewise, people who smoke tobacco are more likely to develop lung cancer, but, in fact, it is possible for non-smokers to develop the disease as well. **Statistics** describes these kinds of situations, where the likelihood of some outcome depends on the starting data.

Descriptive statistics involves analyzing a collection of data to describe its broad properties such average (or mean), what percent of the data falls within a given range, and other such properties. An example of this would be taking all of the test scores from a given class and calculating the average test score. **Inferential statistics** attempts to use data about a subset of some population to make inferences about the rest of the population. An example of this would be taking a collection of students who received tutoring and comparing their results to a collection of students who did not receive tutoring, then using that comparison to try to predict whether the tutoring program in question is beneficial.

To be sure that inferences have a high probability of being true for the whole population, the subset that is analyzed needs to resemble a miniature version of the population as closely as possible. For this reason, statisticians like to choose random samples from the population to study, rather than picking a specific group of people based on some similarity. For example, studying the incomes of people who live in Portland does not tell anything useful about the incomes of people who live in Tallahassee.

A statistical question is answered by collecting data with variability. Data consists of facts and/or statistics (numbers), and variability refers to a tendency to shift or change. Data is a broad term, inclusive of things like height, favorite color, name, salary, temperature, gas mileage, and language. Questions requiring data as an answer are not necessarily statistical questions. If there is no variability in the data, then the question is not statistical in nature. Consider the following examples: what is Mary's favorite color? How much money does your mother make? What was the highest temperature last week? How many miles did your car get on its last tank of gas? How much taller than Bob is Ed?

None of the above are statistical questions because each case lacks variability in the data needed to answer the question. The questions on favorite color, salary, and gas mileage each require a single piece of data, whether a fact or statistic. Therefore, variability is absent. Although the temperature question requires multiple pieces of data (the high temperature for each day), a single, distinct number is the answer. The height question requires two pieces of data, Bob's height and Ed's height, but no difference in variability exists between those two values. Therefore, this is not a statistical question. Statistical questions typically require calculations with data.

Consider the following statistical questions:

How many miles per gallon of gas does the 2016 Honda Civic get? To answer this question, data must be collected. This data should include miles driven and gallons used. Different cars, different drivers, and different driving conditions will produce different results. Therefore, variability exists in the data. To answer the question, the mean (average) value could be determined.

Are American men taller than German men? To answer this question, data must be collected. This data should include the heights of American men and the heights of German men. All American men are not the same height and all German men are not the same height. Some American men are taller than some German men and some German men are taller than some American men. Therefore, variability exists in the data. To answer the question, the median values for each group could be determined and compared.

The following are more examples of statistical questions: What proportion of 4th graders have a favorite color of blue? How much money do teachers make? Is it colder in Boston or Chicago?

An **experiment** is the method in which a hypothesis is tested using a controlled process called the scientific method. A cause and the effect of that cause are measured, and the hypothesis is accepted or rejected. Experiments are usually completed in a controlled environment where the results of a control population are compared to the results of a test population. The groups are selected using a randomization process in which each group has a representative mix of the population being tested. Finally, an **observational study** is similar to an experiment. However, this design is used when circumstances prevent or do not allow for a designated control group and experimental group (e.g., lack of funding or unrealistic expectations). Instead, existing control and test populations must be used, so this method has a lack of randomization.

Statistics involves making decisions and predictions about larger data sets based on smaller data sets. Basically, the information from one part or subset can help predict what happens in the entire data set or population at large. The entire process involves guessing, and the predictions and decisions may not be 100 percent correct all of the time; however, there is some truth to these predictions, and the decisions do have mathematical support. The smaller data set is called a **sample** and the larger data set (in which the decision is being made) is called a **population**. A **random sample** is used as the sample, which is an unbiased collection of data points that represents the population as well as it can. There are many methods of forming a random sample, and all adhere to the fact that every potential data point has a predetermined probability of being chosen.

Describing Distributions

Mean, Median, and Mode
The center of a set of data (statistical values) can be represented by its mean, median, or mode. These are sometimes referred to as measures of central tendency.

Mean
The first property that can be defined for this set of data is the **mean**. This is the same as the average. To find the mean, add up all the data points, then divide by the total number of data points. For example, suppose that in a class of 10 students, the scores on a test were 50, 60, 65, 65, 75, 80, 85, 85, 90, 100. Therefore, the average test score will be:

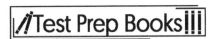

$$\frac{50 + 60 + 65 + 65 + 75 + 80 + 85 + 85 + 90 + 100}{10} = 75.5$$

The mean is a useful number if the distribution of data is normal (more on this later), which roughly means that the frequency of different outcomes has a single peak and is roughly equally distributed on both sides of that peak. However, it is less useful in some cases where the data might be split or where there are some outliers. **Outliers** are data points that are far from the rest of the data. For example, suppose there are 10 executives and 90 employees at a company. The executives make $1000 per hour, and the employees make $10 per hour.

Therefore, the average pay rate will be:

$$\frac{\$1000 \times 10 + \$10 \times 90}{100} = \$109 \ per \ hour$$

In this case, this average is not very descriptive since it's not close to the actual pay of the executives *or* the employees.

Median

Another useful measurement is the **median**. In a data set, the median is the point in the middle. The middle refers to the point where half the data comes before it and half comes after, when the data is recorded in numerical order. For instance, these are the speeds of the fastball of a pitcher during the last inning that he pitched (in order from least to greatest):

$$90, 92, 93, 93, 95, 96, 97, 97, 97$$

There are nine total numbers, so the middle or *median* number is the 5th one, which is 95.

In cases where the number of data points is an even number, then the average of the two middle points is taken. In the previous example of test scores, the two middle points are 75 and 80. Since there is no single point, the average of these two scores needs to be found. The average is:

$$\frac{75 + 80}{2} = 77.5$$

The median is generally a good value to use if there are a few outliers in the data. It prevents those outliers from affecting the "middle" value as much as when using the mean.

Since an outlier is a data point that is far from most of the other data points in a data set, this means an outlier also is any point that is far from the median of the data set. The outliers can have a substantial effect on the mean of a data set, but they usually do not change the median or mode, or do not change them by a large quantity. For example, consider the data set (3, 5, 6, 6, 6, 8). This has a median of 6 and a mode of 6, with a mean of $\frac{34}{6} \approx 5.67$. Now, suppose a new data point of 1000 is added so that the data set is now (3, 5, 6, 6, 6, 8, 1000). The median and mode, which are both still 6, remain unchanged. However, the average is now $\frac{1034}{7}$, which is approximately 147.7. In this case, the median and mode will be better descriptions for most of the data points.

Outliers in a given data set are sometimes the result of an error by the experimenter, but oftentimes, they are perfectly valid data points that must be taken into consideration.

Mode

One additional measure to define for *X* is the **mode**. This is the data point that appears most frequently. If two or more data points all tie for the most frequent appearance, then each of them is considered a mode. In the case of the test scores, where the numbers were 50, 60, 65, 65, 75, 80, 85, 85, 90, 100, there are two modes: 65 and 85.

Quartiles and Percentiles

The **first quartile** of a set of data *X* refers to the largest value from the first ¼ of the data points. In practice, there are sometimes slightly different definitions that can be used, such as the median of the first half of the data points (excluding the median itself if there are an odd number of data points). The term also has a slightly different use: when it is said that a data point lies *in the first quartile*, it means it is less than or equal to the median of the first half of the data points. Conversely, if it lies *at* the first quartile, then it is equal to the first quartile.

When it is said that a data point lies in the **second quartile**, it means it is between the first quartile and the median.

The **third quartile** refers to data that lies between ½ and ¾ of the way through the data set. Again, there are various methods for defining this precisely, but the simplest way is to include all of the data that lie between the median and the median of the top half of the data.

Data that lies in the **fourth quartile** refers to all of the data above the third quartile.

Percentiles may be defined in a similar manner to quartiles. Generally, this is defined in the following manner:

If a data point lies **in the n-th percentile**, this means it lies in the range of the first *n*% of the data.

If a data point lies **at the n-th percentile**, then it means that *n*% of the data lies below this data point.

Standard Deviation

Given a data set *X* consisting of data points $(x_1, x_2, x_3, \ldots x_n)$, the **variance** of *X* is defined to be:

$$\frac{\sum_{i=1}^{n}(x_i - \bar{X})^2}{n}$$

This means that the variance of *X* is the average of the squares of the differences between each data point and the mean of *X*.

Given a data set *X* consisting of data points $(x_1, x_2, x_3, \ldots x_n)$, the **standard deviation** of *X* is defined to be:

$$s_x = \sqrt{\frac{\sum_{i=1}^{n}(x_i - \bar{X})^2}{n}}$$

In other words, the standard deviation is the square root of the variance.

Both the variance and the standard deviation are measures of how much the data tend to be spread out. When the standard deviation is low, the data points are mostly clustered around the mean. When the standard deviation is high, this generally indicates that the data are quite spread out, or else that there are a few substantial outliers.

As a simple example, compute the standard deviation for the data set (1, 3, 3, 5). First, compute the mean, which will be:

$$\frac{1+3+3+5}{4} = \frac{12}{4} = 3$$

Now, find the variance of X with the formula:

$$\sum_{i=1}^{4}(x_i - \bar{X})^2 = (1-3)^2 + (3-3)^2 + (3-3)^2 + (5-3)^2$$

$$-2^2 + 0^2 + 0^2 + 2^2 = 8$$

Therefore, the variance is $\frac{8}{4} = 2$. Taking the square root, the standard deviation will be $\sqrt{2}$.

Note that the standard deviation only depends upon the mean, not upon the median or mode(s). Generally, if there are multiple modes that are far apart from one another, the standard deviation will be high. A high standard deviation does not always mean there are multiple modes, however.

Describing a Set of Data

A set of data can be described in terms of its center, spread, shape and any unusual features. The center of a data set can be measured by its mean, median, or mode. The spread of a data set refers to how far the data points are from the center (mean or median). A data set with all its data points clustered around the center will have a small spread. A data set covering a wide range of values will have a large spread.

When a data set is displayed as a graph like the one below, the shape indicates if a sample is normally distributed, symmetrical, or has measures of skewness. When graphed, a data set with a normal distribution will resemble a bell curve.

If the data set is symmetrical, each half of the graph when divided at the center is a mirror image of the other. If the graph has fewer data points to the right, the data is skewed right. If it has fewer data points to the left, the data is skewed left.

Right-Skewed   Symmetric   Left-Skewed

A description of a data set should include any unusual features such as gaps or outliers. A gap is a span within the range of the data set containing no data points. An outlier is a data point with a value either extremely large or extremely small when compared to the other values in the set.

The graphs above can be referred to as **unimodal** since they all have a single peak. In contrast, a bimodal graph has two peaks.

Geometry Concepts

Properties of Polygons and Circles

A **polygon** is a closed two-dimensional figure consisting of three or more sides. Polygons can be either convex or concave. A polygon that has interior angles all measuring less than 180° is **convex**. A **concave** polygon has one or more interior angles measuring greater than 180°. Examples are shown below.

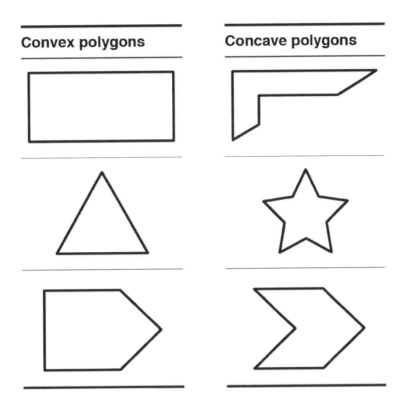

Polygons can be classified by the number of sides (also equal to the number of angles) they have. The following are the names of polygons with a given number of sides or angles:

# of sides	3	4	5	6	7	8	9	10
Name of polygon	Triangle	Quadrilateral	Pentagon	Hexagon	Septagon (or heptagon)	Octagon	Nonagon	Decagon

Equiangular polygons are polygons in which the measure of every interior angle is the same. The sides of equilateral polygons are always the same length. If a polygon is both equiangular and equilateral, the polygon is defined as a **regular polygon**. Examples are shown below.

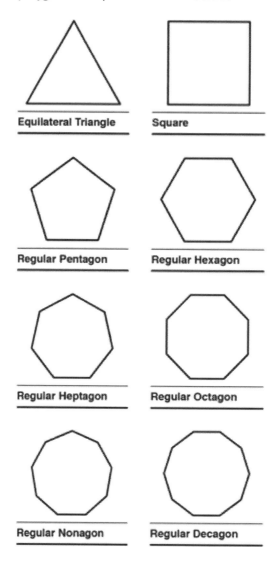

Equilateral Triangle

Square

Regular Pentagon

Regular Hexagon

Regular Heptagon

Regular Octagon

Regular Nonagon

Regular Decagon

Triangles can be further classified by their sides and angles. A triangle with its largest angle measuring 90° is a **right triangle**.

Right triangle

A triangle with the largest angle less than 90° is an **acute triangle**. A triangle with the largest angle greater than 90° is an **obtuse triangle**.

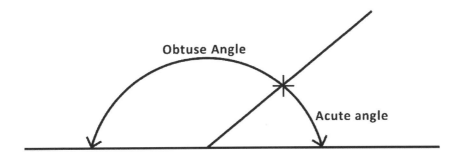

A triangle consisting of two equal sides and two equal angles is an **isosceles triangle**. A triangle with three equal sides and three equal angles is an **equilateral triangle**. A triangle with no equal sides or angles is a **scalene triangle**.

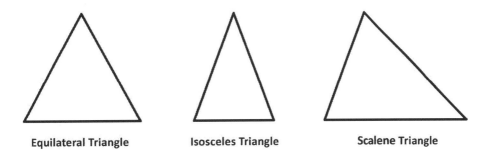

Quadrilaterals can be further classified according to their sides and angles. A quadrilateral with exactly one pair of parallel sides is called a **trapezoid**. A quadrilateral that shows both pairs of opposite sides parallel is a **parallelogram**. Parallelograms include rhombuses, rectangles, and squares. A **rhombus** has four equal sides. A **rectangle** has four equal angles (90° each). A **square** has four 90° angles and four equal sides. Therefore, a square is both a rhombus and a rectangle.

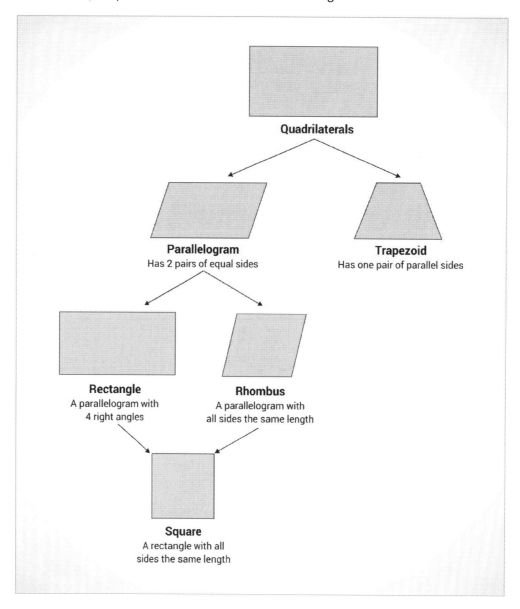

There are many key facts related to geometry that are applicable. The sum of the measures of the angles of a triangle are 180°, and for a quadrilateral, the sum is 360°. Rectangles and squares each have four right angles. A **right angle** has a measure of 90°.

Lines and Angles

In geometry, a **line** connects two points, has no thickness, and extends indefinitely in both directions beyond each point. If the length is finite, it's known as a **line segment** and has two **endpoints**. A **ray** is the straight portion of a line that has one endpoint and extends indefinitely in the other direction. An **angle** is formed when two rays begin at the same endpoint and extend indefinitely. The endpoint of an angle is called a **vertex**. **Adjacent angles** are two side-by-side angles formed from the same ray that have the same endpoint. Angles are measured in **degrees** or **radians**, which is a measure of **rotation**. A

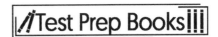

full rotation equals 360 degrees or 2π radians, which represents a circle. Half a rotation equals 180 degrees or π radians and represents a half-circle. Angle measurement is additive. When an angle is broken into two non-overlapping angles, the total measure of the larger angle equals the sum of the two smaller angles. Lines are **coplanar** if they're located in the same plane. Two lines are **parallel** if they are coplanar, extend in the same direction, and never cross. If lines do cross, they're labeled as **intersecting lines** because they "intersect" at one point. If they intersect at more than one point, they're the same line. **Perpendicular lines** are coplanar lines that form a right angle at their point of intersection.

Two lines are parallel if they have the same slope and different intercept. Two lines are perpendicular if the product of their slope equals -1. Parallel lines never intersect unless they are the same line, and perpendicular lines intersect at a right angle. If two lines aren't parallel, they must intersect at one point. Determining equations of lines based on properties of parallel and perpendicular lines appears in word problems. To find an equation of a line, both the slope and a point the line goes through are necessary. Therefore, if an equation of a line is needed that's parallel to a given line and runs through a specified point, the slope of the given line and the point are plugged into the point-slope form of an equation of a line. Secondly, if an equation of a line is needed that's perpendicular to a given line running through a specified point, the negative reciprocal of the slope of the given line and the point are plugged into the point-slope form. Also, if the point of intersection of two lines is known, that point will be used to solve the set of equations. Therefore, to solve a system of equations, the point of intersection must be found. If a set of two equations with two unknown variables has no solution, the lines are parallel.

Classification of Angles

An **angle** consists of two rays that have a common endpoint. This common endpoint is called the **vertex** of the angle. The two rays can be called sides of the angle. The angle below has a vertex at point *B* and the sides consist of ray *BA* and ray *BC*. An angle can be named in three ways:

1. Using the vertex and a point from each side, with the vertex letter in the middle.
2. Using only the vertex. This can only be used if it is the only angle with that vertex.
3. Using a number that is written inside the angle.

The angle below can be written ∠*ABC* (read angle *ABC*), ∠*CBA*, ∠*B*, or ∠1.

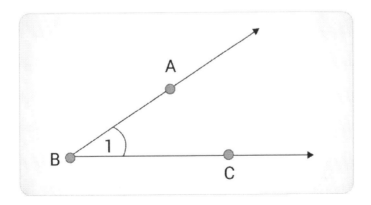

An angle divides a plane, or flat surface, into three parts: the angle itself, the interior (inside) of the angle, and the exterior (outside) of the angle. The figure below shows point *M* on the interior of the angle and point *N* on the exterior of the angle.

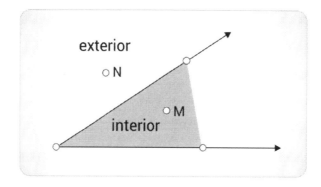

Angles can be measured in units called degrees, with the symbol °. The degree measure of an angle is between 0° and 180° and can be obtained by using a protractor.

A straight angle (or simply a line) measures exactly 180°. A right angle's sides meet at the vertex to create a square corner. A right angle measures exactly 90° and is typically indicated by a box drawn in the interior of the angle. An acute angle has an interior that is narrower than a right angle. The measure of an acute angle is any value less than 90° and greater than 0°. For example, 89.9°, 47°, 12°, and 1°. An obtuse angle has an interior that is wider than a right angle. The measure of an obtuse angle is any value greater than 90° but less than 180°. For example, 90.1°, 110°, 150°, and 179.9°.

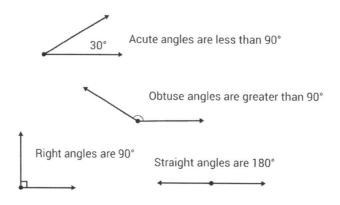

Perimeter and Area

Perimeter is the measurement of a distance around something or the sum of all sides of a polygon. Think of perimeter as the length of the boundary, like a fence. In contrast, **area** is the space occupied by a defined enclosure, like a field enclosed by a fence.

When thinking about perimeter, think about walking around the outside of something. When thinking about area, think about the amount of space or **surface area** something takes up.

Square

The perimeter of a square is measured by adding together all of the sides. Since a square has four equal sides, its perimeter can be calculated by multiplying the length of one side by 4. Thus, the formula is $P = 4 \times s$, where s equals one side. For example, the following square has side lengths of 5 meters:

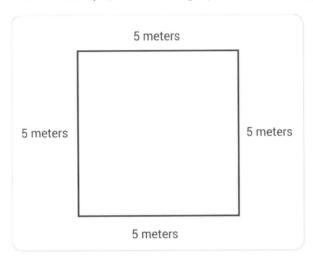

The perimeter is 20 meters because 4 times 5 is 20.

The area of a square is the length of a side squared. For example, if a side of a square is 7 centimeters, then the area is 49 square centimeters. The formula for this example is:

$$A = s^2 = 7^2 = 49 \text{ square centimeters}$$

An example is if the rectangle has a length of 6 inches and a width of 7 inches, then the area is 42 square inches:

$$A = lw = 6(7) = 42 \text{ square inches}$$

Rectangle

Like a square, a rectangle's perimeter is measured by adding together all of the sides. But as the sides are unequal, the formula is different. A rectangle has equal values for its lengths (long sides) and equal values for its widths (short sides), so the perimeter formula for a rectangle is:

$$P = l + l + w + w = 2l + 2w$$

l equals length
w equals width

The area is found by multiplying the length by the width, so the formula is $A = l \times w$.

For example, if the length of a rectangle is 10 inches and the width 8 inches, then the perimeter is 36 inches because:

$$P = 2l + 2w = 2(10) + 2(8) = 20 + 16 = 36 \text{ inches}$$

Triangle

A triangle's perimeter is measured by adding together the three sides, so the formula is $P = a + b + c$, where $a, b,$ and c are the values of the three sides. The area is the product of one-half the base and height so the formula is:

$$A = \frac{1}{2} \times b \times h$$

It can be simplified to:

$$A = \frac{bh}{2}$$

The base is the bottom of the triangle, and the height is the distance from the base to the peak. If a problem asks to calculate the area of a triangle, it will provide the base and height.

For example, if the base of the triangle is 2 feet and the height 4 feet, then the area is 4 square feet. The following equation shows the formula used to calculate the area of the triangle:

$$A = \frac{1}{2} bh = \frac{1}{2}(2)(4) = 4 \text{ square feet}$$

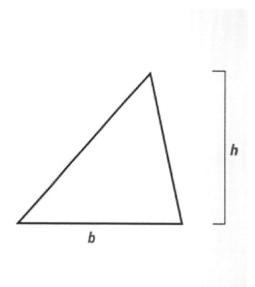

Circle

A circle's perimeter—also known as its circumference—is measured by multiplying the diameter by π.

Diameter is the straight line measured from a point on one side of the circle to a point directly across on the opposite side of the circle.

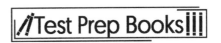

π is referred to as pi and is equal to 3.14 (with rounding).

So the formula is $\pi \times d$.

This is sometimes expressed by the formula $C = 2 \times \pi \times r$, where r is the radius of the circle. These formulas are equivalent, as the radius equals half of the diameter.

The area of a circle is calculated through the formula $A = \pi \times r^2$. The test will indicate either to leave the answer with π attached or to calculate to the nearest decimal place, which means multiplying by 3.14 for π.

The **arc of a circle** is the distance between two points on the circle. The length of the arc of a circle in terms of **degrees** is easily determined if the value of the central angle is known. The length of the arc is simply the value of the central angle. In this example, the length of the arc of the circle in degrees is 75°.

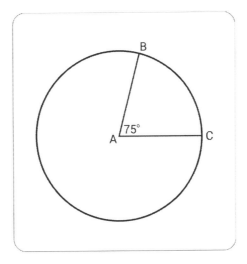

To determine the length of the arc of a circle in distance, the values for both the central angle and the radius must be known. This formula is:

$$\frac{central\ angle}{360°} = \frac{arc\ length}{2\pi r}$$

The equation is simplified by cross-multiplying to solve for the arc length.

In the following example, to solve for arc length, substitute the values of the central angle (75°) and the radius (10 inches) into the equation above.

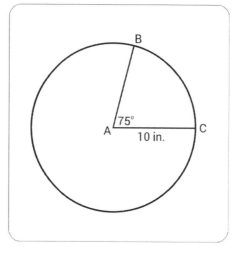

$$\frac{75°}{360°} = \frac{\text{arc length}}{2(3.14)(10\text{in.})}$$

To solve the equation, first cross-multiply: 4710 = 360(arc length). Next, divide each side of the equation by 360. The result of the formula is that the arc length is 13.1 (rounded).

Irregular Shapes

The perimeter of an irregular polygon is found by adding the lengths of all of the sides. In cases where all of the sides are given, this will be very straightforward, as it will simply involve finding the sum of the provided lengths. Other times, a side length may be missing and must be determined before the perimeter can be calculated. Consider the example below:

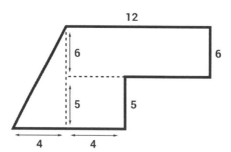

All of the side lengths are provided except for the angled side on the left. Test takers should notice that this is the hypotenuse of a right triangle. The other two sides of the triangle are provided (the base is 4 and the height is 6 + 5 = 11). The Pythagorean Theorem can be used to find the length of the hypotenuse, remembering that $a^2 + b^2 = c^2$.

Substituting the side values provided yields $(4)^2 + (11)^2 = c^2$.

Therefore, $c = \sqrt{16 + 121} = 11.7$

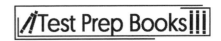

Finally, the perimeter can be found by adding this new side length with the other provided lengths to get the total length around the figure:

$$4+4+5+8+6+12+11.7=50.7$$

Although units are not provided in this figure, remember that reporting units with a measurement is important.

The area of an irregular polygon is found by decomposing, or breaking apart, the figure into smaller shapes. When the area of the smaller shapes is determined, these areas are added together to produce the total area of the area of the original figure. Consider the same example provided before:

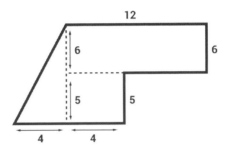

The irregular polygon is decomposed into two rectangles and a triangle. The area of the large rectangles ($A = l \times w \rightarrow A = 12 \times 6$) is 72 square units. The area of the small rectangle is 20 square units:

$$(A = 4 \times 5)$$

The area of the triangle ($A = \frac{1}{2} \times b \times h \rightarrow A = \frac{1}{2} \times 4 \times 11$) is 22 square units. The sum of the areas of these figures produces the total area of the original polygon:

$$A = 72 + 20 + 22 \rightarrow A = 114 \text{ square units}$$

Volume and Surface Area of Three-Dimensional Shapes

Geometry in three dimensions is similar to geometry in two dimensions. The main new feature is that three points now define a unique plane that passes through each of them. Three-dimensional objects can be made by putting together two-dimensional figures in different surfaces. Below, some of the possible three-dimensional figures will be provided, along with formulas for their volumes and surface areas.

Volume is the measurement of how much space an object occupies, like how much space is in the cube. Volume questions will ask how much of something is needed to completely fill the object. The most common surface area and volume questions deal with spheres, cubes, and rectangular prisms.

Surface area of a three-dimensional figure refers to the number of square units needed to cover the entire surface of the figure. This concept is similar to using wrapping paper to completely cover the outside of a box. For example, if a triangular pyramid has a surface area of 17 square inches (written $17in^2$), it will take 17 squares, each with sides one inch in length, to cover the entire surface of the pyramid. Surface area is also measured in square units.

A **rectangular prism** is a box whose sides are all rectangles meeting at 90° angles. Such a box has three dimensions: length, width, and height. If the length is x, the width is y, and the height is z, then the volume is given by $V = xyz$.

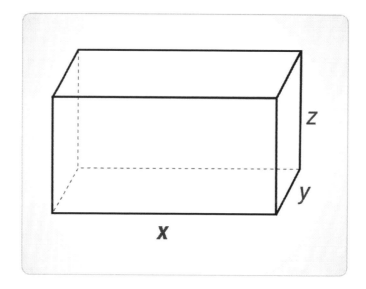

The surface area will be given by computing the surface area of each rectangle and adding them together. There is a total of six rectangles. Two of them have sides of length x and y, two have sides of length y and z, and two have sides of length x and z. Therefore, the total surface area will be given by:

$$SA = 2xy + 2yz + 2xz$$

A **cube** is a special type of rectangular solid in which its length, width, and height are the same. If this length is s, then the formula for the volume of a cube is $V = s \times s \times s$. The surface area of a cube is $SA = 6s^2$.

A **rectangular pyramid** is a figure with a rectangular base and four triangular sides that meet at a single vertex. If the rectangle has sides of length x and y, then the volume will be given by $V = \frac{1}{3}xyh$.

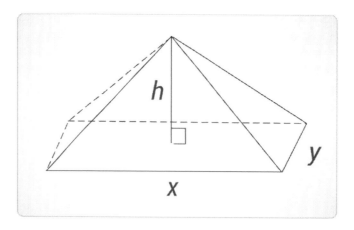

Many three-dimensional figures (solid figures) can be represented by nets consisting of rectangles and triangles. The surface area of such solids can be determined by adding the areas of each of its faces and

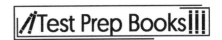

bases. Finding the surface area using this method requires calculating the areas of rectangles and triangles. To find the area (A) of a rectangle, the length (l) is multiplied by the width (w) → $A = l \times w$. The area of a rectangle with a length of 8cm and a width of 4cm is calculated:

$$A = (8cm) \times (4cm) \rightarrow A = 32cm^2$$

To calculate the area (A) of a triangle, the product of $\frac{1}{2}$, the base (b), and the height (h) is found:

$$A = \frac{1}{2} \times b \times h$$

Note that the height of a triangle is measured from the base to the vertex opposite of it forming a right angle with the base. The area of a triangle with a base of 11cm and a height of 6cm is calculated:

$$A = \frac{1}{2} \times (11cm) \times (6cm)$$

$$A = 33cm^2$$

Consider the following triangular prism, which is represented by a net consisting of two triangles and three rectangles.

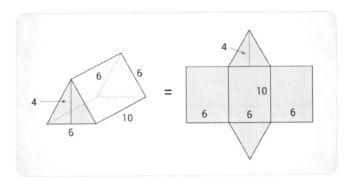

The surface area of the prism can be determined by adding the areas of each of its faces and bases. The surface area (SA) = area of triangle + area of triangle + area of rectangle + area of rectangle + area of rectangle.

$$SA = \left(\frac{1}{2} \times b \times h\right) + \left(\frac{1}{2} \times b \times h\right) + (l \times w) + (l \times w) + (l \times w)$$

$$SA = \left(\frac{1}{2} \times 6 \times 4\right) + \left(\frac{1}{2} \times 6 \times 4\right) + (6 \times 10) + (6 \times 10) + (6 \times 10)$$

$$SA = (12) + (12) + (60) + (60) + (60)$$

$$SA = 204 \; square \; units$$

A **sphere** is a set of points all of which are equidistant from some central point. It is like a circle, but in three dimensions. The volume of a sphere of radius r is given by:

$$V = \frac{4}{3}\pi r^3$$

The surface area is given by $A = 4\pi r^2$.

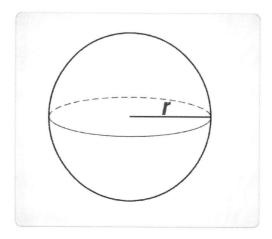

The volume of a **cylinder** is then found by adding a third dimension onto the circle. Volume of a cylinder is calculated by multiplying the area of the base (which is a circle) by the height of the cylinder. Doing so results in the equation $V = \pi r^2 h$. The volume of a **cone** is $\frac{1}{3}$ of the volume of a cylinder. Therefore, the formula for the volume of a cone is:

$$\frac{1}{3}\pi r^2 h$$

Plane Geometry

Algebraic equations can be used to describe geometric figures in the plane. The method for doing so is to use the **Cartesian coordinate plane**. The idea behind these Cartesian coordinates (named for mathematician and philosopher Descartes) is that from a specific point on the plane, known as the **origin**, one can specify any other point by saying *how far to the right or left* and *how far up or down*.

The plane is covered with a grid. The two directions, right to left and bottom to top, are called **axes** (singular **axis**). When working with x and y variables, the x variable corresponds to the right and left axis, and the y variable corresponds to the up and down axis.

Any point on the grid is found by specifying how far to travel from the center along the x-axis and how far to travel along the y-axis. The ordered pair can be written as (x, y). A positive x value means go to the right on the *x*-axis, while a negative *x* value means to go to the left. A positive *y* value means to go up, while a negative value means to go down. Several points are shown as examples in the figure.

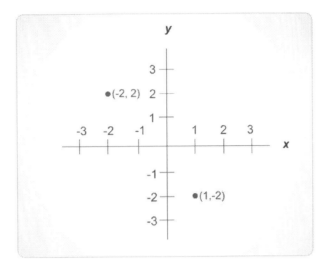

Cartesian Coordinate Plane

The coordinate plane can be divided into four **quadrants**. The upper-right part of the plane is called the first quadrant, where both *x* and *y* are positive. The second quadrant is the upper-left, where x is negative but y is positive. The third quadrant is the lower left, where both x and y are negative. Finally, the fourth quadrant is in the lower right, where x is positive but y is negative. These quadrants are often written with Roman numerals:

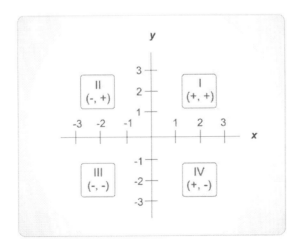

In addition to graphing individual points as shown above, the graph lines and curves in the plane can be graphed corresponding to equations. In general, if there is some equation involving *x* and *y*, then the graph of that equation consists of all the points (x, y) in the Cartesian coordinate plane, which satisfy this equation.

Given the equation $y = x + 2$, the point $(0, 2)$ is in the graph, since $2 = 0 + 2$ is a true equation. However, the point $(1, 4)$ will *not* be in the graph, because $4 = 1 + 2$ is false.

Proving Theorems with Coordinates

Many important formulas and equations exist in geometry that use coordinates. The distance between two points (x_1, y_1) and (x_2, y_2) is:

$$d = \sqrt{(x_2 - x_1)^2 + (y_2 - y_1)^2}$$

The slope of the line containing the same two points is $m = \frac{y_2 - y_1}{x_2 - x_1}$. Also, the midpoint of the line segment with endpoints (x_1, y_1) and (x_2, y_2) is:

$$M = \left(\frac{x_1 + x_2}{2}, \frac{y_1 + y_2}{2}\right)$$

The equations of a circle, parabola, ellipse, and hyperbola can also be used to prove theorems algebraically. Knowing when to use which formula or equation is extremely important, and knowing which formula applies to which property of a given geometric shape is an integral part of the process. In some cases, there are a number of ways to prove a theorem; however, only one way is required.

Formulas for Ratios

If a line segment with endpoints (x_1, y_1) and (x_2, y_2) is partitioned into two equal parts, the formula for midpoint is used. Recall this formula is $M = \left(\frac{x_1 + x_2}{2}, \frac{y_1 + y_2}{2}\right)$ and the ratio of line segments is 1:1. However, if the ratio needs to be anything other than 1:1, a different formula must be used. Consider a ratio that is $a:b$. This means the desired point that partitions the line segment is $\frac{a}{a+b}$ of the way from (x_1, y_1) to (x_2, y_2). The actual formula for the coordinate is:

$$\left(\frac{bx_1 + ax_2}{a + b}, \frac{by_1 + ay_2}{a + b}\right)$$

Computing Side Length, Perimeter, and Area

The side lengths of each shape can be found by plugging the endpoints into the distance formula:

$$d = \sqrt{(x_2 - x_1)^2 + (y_2 - y_1)^2}$$

between two ordered pairs (x_1, y_1) and (x_2, y_2). The distance formula is derived from the Pythagorean theorem. Once the side lengths are found, they can be added together to obtain the perimeter of the given polygon. Simplifications can be made for specific shapes such as squares and equilateral triangles. For example, one side length can be multiplied by 4 to obtain the perimeter of a square. Also, one side length can be multiplied by 3 to obtain the perimeter of an equilateral triangle. A similar technique can be used to calculate areas. For polygons, both side length and height can be found by using the same distance formula. Areas of triangles and quadrilaterals are straightforward through the use of $A = \frac{1}{2}bh$ or $A = bh$, depending on the shape. To find the area of other polygons, their shapes can be partitioned into rectangles and triangles. The areas of these simpler shapes can be calculated and then added together to find the total area of the polygon.

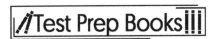

Transformations

Given a figure drawn on a plane, many changes can be made to that figure, including rotation, translation, and reflection. **Rotations** turn the figure about a point, **translations** slide the figure, and **reflections** flip the figure over a specified line. When performing these transformations, the original figure is called the **pre-image**, and the figure after transformation is called the **image**.

More specifically, **translation** means that all points in the figure are moved in the same direction by the same distance. In other words, the figure is slid in some fixed direction. Of course, while the entire figure is slid by the same distance, this does not change any of the measurements of the figures involved. The result will have the same distances and angles as the original figure.

In terms of Cartesian coordinates, a translation means a shift of each of the original points (x, y) by a fixed amount in the x and y directions, to become $(x + a, y + b)$.

Another procedure that can be performed is called **reflection**. To do this, a line in the plane is specified, called the **line of reflection**. Then, take each point and flip it over the line so that it is the same distance from the line but on the opposite side of it. This does not change any of the distances or angles involved, but it does reverse the order in which everything appears.

To reflect something over the x-axis, the points (x, y) are sent to $(x, -y)$. To reflect something over the y-axis, the points (x, y) are sent to the points $(-x, y)$. Flipping over other lines is not something easy to express in Cartesian coordinates. However, by drawing the figure and the line of reflection, the distance to the line and the original points can be used to find the reflected figure.

Example: Reflect this triangle with vertices (-1, 0), (2, 1), and (2, 0) over the y-axis. The pre-image is shown below.

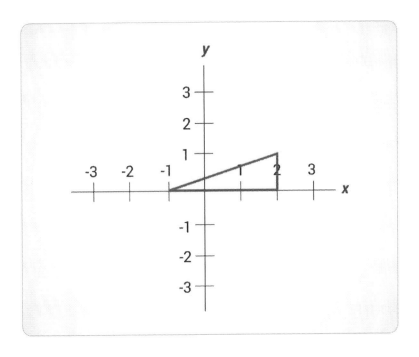

To do this, flip the *x* values of the points involved to the negatives of themselves, while keeping the *y* values the same. The image is shown here.

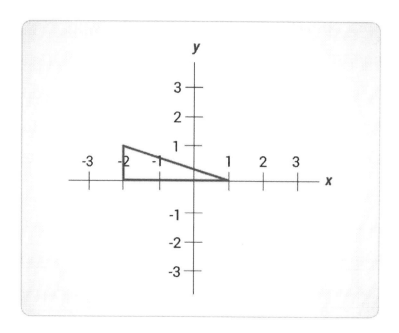

The new vertices will be (1, 0), (-2, 1), and (-2, 0).

Another procedure that does not change the distances and angles in a figure is **rotation**. In this procedure, pick a center point, then rotate every vertex along a circle around that point by the same angle. This procedure is also not easy to express in Cartesian coordinates, and this is not a requirement on this test. However, as with reflections, it's helpful to draw the figures and see what the result of the rotation would look like. This transformation can be performed using a compass and protractor.

Each one of these transformations can be performed on the coordinate plane without changes to the original dimensions or angles.

If two figures in the plane involve the same distances and angles, they are called **congruent figures**. In other words, two figures are congruent when they go from one form to another through reflection, rotation, and translation, or a combination of these.

Remember that rotation and translation will give back a new figure that is identical to the original figure, but reflection will give back a mirror image of it.

To recognize that a figure has undergone a rotation, check to see that the figure has not been changed into a mirror image, but that its orientation has changed (that is, whether the parts of the figure now form different angles with the *x* and *y* axes).

To recognize that a figure has undergone a translation, check to see that the figure has not been changed into a mirror image, and that the orientation remains the same.

To recognize that a figure has undergone a reflection, check to see that the new figure is a mirror image of the old figure.

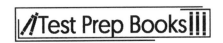

Keep in mind that sometimes a combination of translations, reflections, and rotations may be performed on a figure.

Dilation

A **dilation** is a transformation that preserves angles, but not distances. This can be thought of as stretching or shrinking a figure. If a dilation makes figures larger, it is called an **enlargement**. If a dilation makes figures smaller, it is called a **reduction**. The easiest example is to dilate around the origin. In this case, multiply the x and y coordinates by a **scale factor**, k, sending points (x, y) to (kx, ky).

As an example, draw a dilation of the following triangle, whose vertices will be the points (-1, 0), (1, 0), and (1, 1).

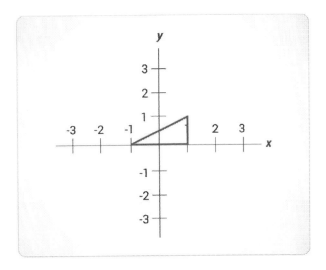

For this problem, dilate by a scale factor of 2, so the new vertices will be (-2, 0), (2, 0), and (2, 2).

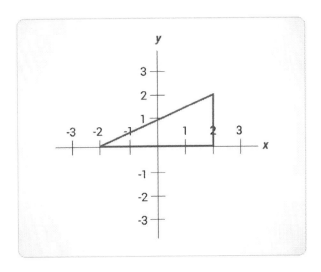

Note that after a dilation, the distances between the vertices of the figure will have changed, but the angles remain the same. The two figures that are obtained by dilation, along with possibly translation, rotation, and reflection, are all *similar* to one another. Another way to think of this is that similar figures have the same number of vertices and edges, and their angles are all the same. Similar figures have the same basic shape but are different in size.

Symmetry

Using the types of transformations above, if an object can undergo these changes and not appear to have changed, then the figure is symmetrical. If an object can be split in half by a line and flipped over that line to lie directly on top of itself, it is said to have **line symmetry**. An example of both types of figures is seen below.

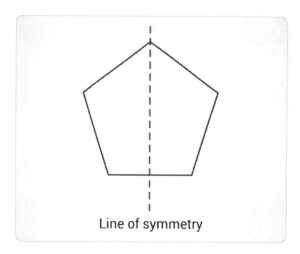

Line of symmetry

If an object can be rotated about its center to any degree smaller than 360, and it lies directly on top of itself, the object is said to have **rotational symmetry**. An example of this type of symmetry is shown below. The pentagon has an order of 5.

The rotational symmetry lines in the figure above can be used to find the angles formed at the center of the pentagon. Knowing that all of the angles together form a full circle, at 360 degrees, the figure can be split into 5 angles equally. By dividing the 360° by 5, each angle is 72°.

Given the length of one side of the figure, the perimeter of the pentagon can also be found using rotational symmetry. If one side length was 3 cm, that side length can be rotated onto each other side length four times. This would give a total of 5 side lengths equal to 3 cm. To find the perimeter, or distance around the figure, multiply 3 by 5. The perimeter of the figure would be 15 cm.

If a line cannot be drawn anywhere on the object to flip the figure onto itself or rotated less than or equal to 180 degrees to lay on top of itself, the object is asymmetrical. Examples of these types of figures are shown below.

Symmetric   Asymmetric

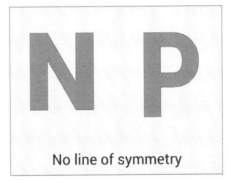

No line of symmetry

Right Triangles: Pythagorean Theorem and Trigonometric Ratio

The value of a missing side of a right triangle may be determined two ways. The first way is to apply the Pythagorean Theorem, and the second way is to apply Trigonometric Ratios. The Pythagorean Theorem states that for every right triangle, the square of the length of the hypotenuse is equal to the sum of the squares of the lengths of the remaining two sides. The hypotenuse is the longest side of a right triangle and is also the side opposite the right angle.

According to the diagram, $a^2 + b^2 = c^2$, where c represents the hypotenuse, and a and b represent the lengths of the remaining two sides of the right triangle.

The Pythagorean Theorem may be applied a multitude of ways. For example, a person wishes to build a garden in the shape of a rectangle, having the dimensions of 5 feet by 8 feet. The garden's design includes a diagonal board to separate various types of plants. The Pythagorean Theorem can be used to determine the length of the diagonal board.

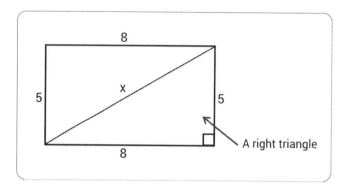

$$a^2 + b^2 = c^2$$

$$5^2 + 8^2 = c^2$$

$$25 + 64 = c^2$$

$$c = \sqrt{89}$$

$$c = 9.43$$

To solve for unknown sides of a right triangle using trigonometric ratios, the sine, cosine, and tangent are required.

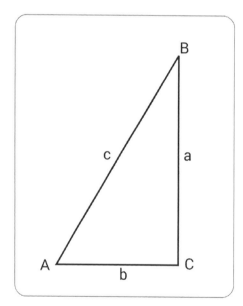

In the image above, angles are denoted by capital letters, and sides are denoted by lowercase letters. When examining angle A, b is the adjacent side, a is the opposite side, and c is the hypotenuse side. The various ratios of the lengths of the sides of the right triangle are used to find the sine, cosine, and tangent of angle A.

Thus:

$$\sin(A) = \frac{opposite}{hypotenuse}$$

$$\cos(A) = \frac{adjacent}{hypotenuse}$$

$$\tan(A) = \frac{opposite}{adjacent}$$

After substituting variables for the sides of the right triangle, $\sin(A) = \frac{a}{c}$, $\cos(A) = \frac{b}{c}$, and $\tan(A) = \frac{a}{b}$.

As a real-world example, the height of a tree can be discovered by using the information above. Surveying equipment can determine the tree's angle of inclination is 55.3 degrees, and the distance from the tree is 10 feet.

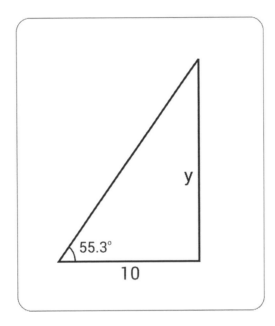

To find the height of the tree, substitute the known values into the trigonometric ratio of tangent:

$$\tan(55.3) = \frac{y}{10}$$

$$10 \times \tan(55.3) = y$$

$$10 \times 1.44418 = y$$

$$y = 14.4418$$

Similarity, Congruence, and Triangles

Triangles are similar if they have the same shape, the same angle measurements, and their sides are proportional to one another. Triangles are congruent if the angles of the triangles are equal in measurement and the sides of the triangles are equal in measurement.

There are five ways to show that a triangle is congruent.

- SSS (Side-Side-Side Postulate): When all three corresponding sides are equal in length, then the two triangles are congruent.

- SAS (Side-Angle-Side Postulate): If a pair of corresponding sides and the angle in between those two sides are equal, then the two triangles are congruent.

- ASA (Angle-Side-Angle Postulate): If a pair of corresponding angles are equal and the side within those angles are equal, then the two triangles are equal.

- AAS (Angle-Angle-Side Postulate): When a pair of corresponding angles for two triangles and a non-included side are equal, then the two triangles are congruent.

- HL (Hypotenuse-Leg Theorem): If two right triangles have the same hypotenuse length, and one of the other sides are also the same length, then the two triangles are congruent.

If two triangles are discovered to be similar or congruent, this information can assist in determining unknown parts of triangles, such as missing angles and sides.

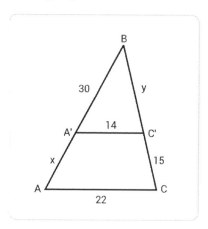

In the triangle shown above, AC and $A'C'$ are parallel lines. Therefore, BA is a transversal that intersects the two parallel lines. The corresponding angles $BA'C'$ and BAC are congruent. In a similar way, BC is also a transversal. Therefore, angle $BC'A'$ and BCA are congruent. If two triangles have two congruent angles, the triangles are similar. If the triangles are similar, their corresponding sides are proportional.

Therefore, the following equation is established:

$$\frac{30 + x}{30} = \frac{22}{14} = \frac{y + 15}{y}$$

$$\frac{30 + x}{30} = \frac{22}{14}$$

$$x = 17.1$$

$$\frac{22}{14} = \frac{y + 15}{y}$$

$$y = 26.25$$

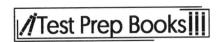

The example below involves the question of congruent triangles. The first step is to examine whether the triangles are congruent. If the triangles are congruent, then the measure of a missing angle can be found.

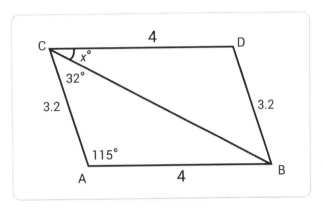

The above diagram provides values for angle measurements and side lengths in triangles *CAB* and *CDB*. Note that side *CA* is 3.2 and side *DB* is 3.2. Side *CD* is 4 and side *AB* is 4. Furthermore, line *CB* is congruent to itself by the reflexive property. Therefore, the two triangles are congruent by SSS (Side-Side-Side). Because the two triangles are congruent, all of the corresponding parts of the triangles are also congruent. Therefore, angle *x* is congruent to the inside of the angle for which a measurement is not provided in Triangle *CAB*. Thus, $115° + 32° = 147°$. A triangle measures 180°, therefore $180° - 147° = 33°$. *Angle x* = 33°, because the two triangles are reversed.

Complementary Angle Theorem

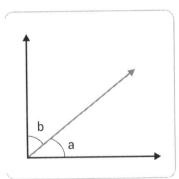

Two angles are complementary if the sum of the two angles equals 90°.

In the above diagram $Angle\ a + Angle\ b = 90°$. Therefore, the two angles are complementary. Certain trigonometric rules are also associated with complementary angles.

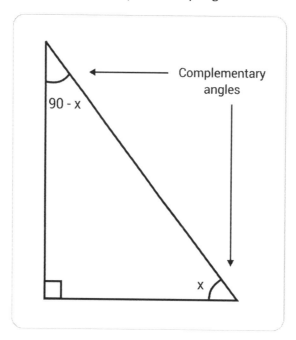

In the diagram above of a right triangle, if Angle *A* and Angle *C* are determined to be complementary angles, then certain relationships can be stated between the trigonometry of those angles.

$$sin(90° - x) = cos\ x$$

$$cos(90° - x) = sin\ x$$

For example, the sine of 80 degrees equals the cosine of $(90° - 80°)$, which is the cos $(10°)$.

This is true because the sine of an angle in a right triangle is equal to the cosine of its complement. Sine is known as the conjunction of cosine, and cosine is known as the conjunction of sine.

Examples:

1. $cos5° = sin\ x°$?
2. $sin(90° - x) = ?\ sin(90° - x) =?$

For problem number 1, the student should remember that $sin(90° - x) = cos\ x$. Cos 5° would be the same as $sin(90 - 5)°$. Therefore, $cos5° = sin85°$.

For problem number 2, the student would use the same fact that $sin(90° - x)° = cos\ x$.

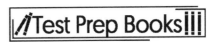

An *acute angle* is an angle that is less than 90°. If Angle *A* and Angle *B* are acute angles of a right triangle, then $sinA = cosB$. Therefore, the sine of any acute angle in a right triangle is equal to the cosine of its complement, and the cosine of any acute angle is equal to the sine of its complement.

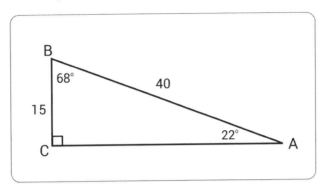

The example above is a right triangle. If only the value of angle *BAC* (which is 22°) was provided, the student would be able to figure out the value for angle *CBA* (68°) by knowing that a triangle is made up of 180°($180° - 90° - 22° = 68°$). From the information given about acute angles on the previous page, the following statement is true:

Sine (angle *BAC*) = $\frac{15}{40}$, which is equivalent to the Cos (angle *CBA*) = $\frac{15}{40}$

Circles on the Coordinate Plane

If a circle is placed on the coordinate plane with the center of the circle at the origin (0,0), then point (*x*, *y*) is a point on the circle. Furthermore, the line extending from the center to point (*x*, *y*) is the radius, or *r*. By applying the Pythagorean Theorem ($a^2 + b^2 = c^2$), it can be stated that $x^2 + y^2 = r^2$. However, the center of the circle does not always need to be on the origin of the coordinate plane.

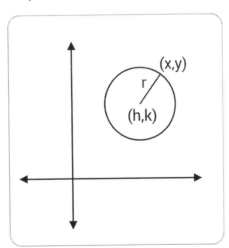

In the diagram above, the center of the circle is noted by (*h*, *k*). By applying the distance formula, the equation becomes: $= \sqrt{(x - h)^2 + (y - k)^2}$. When squaring both sides of the equation, the result is the standard form of a circle with the center (*h*, *k*) and radius *r*. Namely, $r^2 = (x - h)^2 + (y - k)^2$, where *r* = radius and center = (*h*, *k*). The following examples may be solved by using this information:

Example: Graph the equation $-x^2 + y^2 = 25$

To graph this equation, first note that the center of the circle is (0, 0). The radius is the positive square root of 25 or 5.

Example: Find the equation for the circle below.

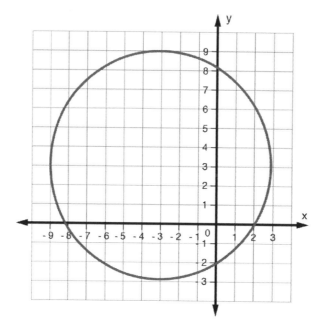

To find the equation for the circle, note that its center is not zero. Therefore, to find the circle's center, draw vertical and horizontal diameters to examine where they intersect. The center is located at point: (-3, 3). Next, count the number of spaces from the center to the outside of the circle. This number is 6. Therefore, 6 is the radius. Finally, plug in the numbers that are known into the standard equation for a circle:

$$36 = \left(x - (-3)\right)^2 + (y - 3)^2$$

$$or$$

$$36 = (x + 3)^2 + (y - 3)^2$$

It is possible to determine whether a point lies on a circle or not within the coordinate plane. For example, a circle has a center of (2, -5), and a radius of 6 centimeters. The first step is to apply the equation of a circle, which is $r^2 = (x - h)^2 + (y - k)^2$, where r = radius and the center = (h, k). Next, substitute the numbers for the center point and the number for the radius. This action simplifies the equation to $36 = (x - 2)^2 + (y + 5)^2$. Note that the radius of 6 was squared to get 36.

To prove that the point (2, -1) lies on the circle, apply the equation of the circle that was just used and input the values of (2, -1) for x and y in the equation.

$$36 = (x - 2)^2 + (y + 5)^2$$

$$36 = (2 - 2)^2 + (1 + 5)^2$$

$$36 = (0)^2 + (6)^2$$

$$36 = 36$$

Because the left side of the equation equals the right side of the equation, point (2, 1) lies on the given circle.

Practice Questions

1. The graph of which function has an x-intercept of -2?
    a. $y = 2x - 3$
    b. $y = 4x + 2$
    c. $y = x^2 + 5x + 6$
    d. $y = -\frac{1}{2} \times 2^x$

2. The table below displays the number of three-year-olds at Kids First Daycare who are potty-trained and those who still wear diapers.

	Potty-trained	Wear diapers	
Boys	26	22	48
Girls	34	18	52
	60	40	

What is the probability that a three-year-old girl chosen at random from the school is potty-trained?
    a. 52 percent
    b. 34 percent
    c. 65 percent
    d. 57 percent

3. What is the volume of a cylinder, in terms of π, with a radius of 5 inches and a height of 10 inches?
    a. $250\,\pi$ in³
    b. $50\,\pi$ in³
    c. $100\,\pi$ in³
    d. $200\,\pi$ in³

4. What is the function that forms an equivalent graph to $y = \cos(x)$?
    a. $y = \tan(x)$
    b. $y = \csc(x)$
    c. $y = \sin\left(x + \frac{\pi}{2}\right)$
    d. $y = \sin\left(x - \frac{\pi}{2}\right)$

5. A rectangle was formed out of pipe cleaner. Its length was $\frac{1}{2}$ feet and its width was $\frac{11}{2}$ inches. What was its area in square inches?
    a. $\frac{11}{4}$ inch²
    b. $\frac{11}{2}$ inch²
    c. 22 inch²
    d. 33 inch²

Answer Explanations

1. C: An x-intercept is the point where the graph crosses the x-axis. At this point, the value of y is 0. To determine if an equation has an x-intercept of -2, substitute -2 for x, and calculate the value of y. If the value of -2 for x corresponds with a y-value of 0, then the equation has an x-intercept of -2. The only answer choice that produces this result is Choice C:

$$0 = (-2)^2 + 5(-2) + 6$$

2. C: The conditional frequency of a girl being potty-trained is calculated by dividing the number of potty-trained girls by the total number of girls: $34 \div 52 = 0.65$. To determine the conditional probability, multiply the conditional frequency by 100: $0.65 \times 100 = 65\%$.

3. A: The volume of a cylinder is $\pi r^2 h$, and $\pi \times 5^2 \times 10$ is $250 \pi \ in^3$. Choice B is not the correct answer because that is $5^2 \times 2\pi$. Choice C is not the correct answer since that is $5in \times 10\pi$. Choice D is not the correct answer because that is $10^2 \times 2in$.

4. C: Graphing the function $y = \cos(x)$ shows that the curve starts at $(0, 1)$, has an amplitude of 2, and a period of 2π. This same curve can be constructed using the sine graph, by shifting the graph to the left $\frac{\pi}{2}$ units. This equation is in the form:

$$y = \sin(x + \frac{\pi}{2})$$

5. D: Area = length x width. The answer must be in square inches, so all values must be converted to inches. $\frac{1}{2}$ ft is equal to 6 inches. Therefore, the area of the rectangle is equal to $6 \times \frac{11}{2} = \frac{66}{2} = 33$ square inches.

Advanced Algebra and Functions

Linear Equations

Solving Linear Equations

When asked to solve a linear equation, it requires determining a numerical value for the unknown variable. Given a linear equation involving addition, subtraction, multiplication, and division, isolation of the variable is done by working backward. Addition and subtraction are inverse operations, as are multiplication and division; therefore, they can be used to cancel each other out.

The first steps to solving linear equations are to distribute if necessary and combine any like terms that are on the same side of the equation. Sides of an equation are separated by an = sign. Next, the equation should be manipulated to get the variable on one side. Whatever is done to one side of an equation, must be done to the other side to remain equal. Then, the variable should be isolated by using inverse operations to undo the order of operations backward. Undo addition and subtraction, then undo multiplication and division. For example:

Solve $4(t - 2) + 2t - 4 = 2(9 - 2t)$

Distribute: $4t - 8 + 2t - 4 = 18 - 4t$

Combine like terms: $6t - 12 = 18 - 4t$

Add 4t to each side to move the variable: $10t - 12 = 18$

Add 12 to each side to isolate the variable: $10t = 30$

Divide each side by 10 to isolate the variable: $t = 3$

The answer can be checked by substituting the value for the variable into the original equation and ensuring both sides calculate to be equal.

Algebraically Solving Linear Equations (or Inequalities) in One Variable

Linear equations in one variable and linear inequalities in one variable can be solved following similar processes. Although they typically have one solution, a linear equation can have no solution or can have a solution set of all real numbers. Solution sets for linear inequalities typically consist of an infinite number of values either greater or less than a given value (where the given value may or may not be included in the set). However, a linear inequality can have no solution or can have a solution set consisting of all real numbers.

Linear Equations in One Variable – Special Cases

Solving a linear equation produces a value for the variable that makes the algebraic statement true. If there is no value for the variable that would make the statement true, there is no solution to the equation. Here's a sample equation: $x + 3 = x - 1$. There is no value for x in which adding 3 to the value would produce the same result as subtracting 1 from that value. Conversely, if any value for the variable would make a true statement, the equation has an infinite number of solutions. Here's another

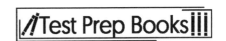

sample equation: $3x + 6 = 3(x + 2)$. Any real number substituted for x would result in a true statement (both sides of the equation are equal).

By manipulating equations similar to the two above, the variable of the equation will cancel out completely. If the constants that are left express a true statement (ex., $6 = 6$), then all real numbers are solutions to the equation. If the constants left express a false statement (ex., $3 = -1$), then there is no solution to the equation.

A question on this material may present a linear equation with an unknown value for either a constant or a coefficient of the variable and ask to determine the value that produces an equation with no solution or infinite solutions. For example:

$3x + 7 = 3x + 10 + n$; Find the value of n that would create an equation with an infinite number of solutions for the variable x.

To solve this problem, the equation should be manipulated so the variable x will cancel. To do this, $3x$ should be subtracted from both sides, which would leave $7 = 10 + n$. By subtracting 10 on both sides, it is determined that $n = -3$. Therefore, a value of -3 for n would result in an equation with a solution set of all real numbers.

If the same problem asked for the equation to have no solution, the value of n would be all real numbers except -3.

Linear Inequalities in One Variable – Special Cases
A linear inequality can have a solution set consisting of all real numbers or can contain no solution. When solved algebraically, a linear inequality in which the variable cancels out and results in a true statement (ex., $7 \geq 2$) has a solution set of all real numbers. A linear inequality in which the variable cancels out and results in a false statement (ex., $7 \leq 2$) has no solution.

Compound Inequalities
A compound inequality is a pair of inequalities joined by *and* or *or*. Given a compound inequality, to determine its solution set, both inequalities should be solved for the given variable. The solution set for a compound inequality containing *and* consists of all the values for the variable that make both inequalities true. If solving the compound inequality results in $x > -9$ and $x \leq 6$, the solution set would consist of all values between -2 and 3, including 3. This may also be written as follows: $-9 < x \leq 6$. Due to the graphs of their solution sets (shown below), compound inequalities such as these are referred to as conjunctions.

If there are no values that would make both inequalities of a compound inequality containing *and* true, then there is no solution. An example would be $x > 2$ and $x \leq 0$.

The solution set for a compound inequality containing *or* consists of all the values for the variable that make at least one of the inequalities true. The solution set for the compound inequality $x \leq -2$ or $x > 1$ consists of all values less than 2, 6, and all values greater than 6. Due to the graphs of their solution sets (shown below), compound inequalities such as these are referred to as disjunctions.

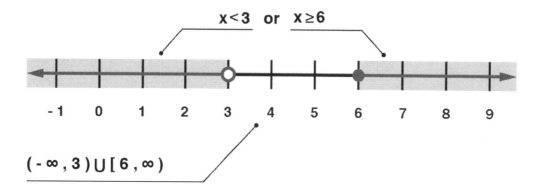

If the two inequalities for a compound inequality containing *or* "overlap," then the solution set contains all real numbers. An example would be $x > 2$ or $x < 7$. Any number would make at least one of these true.

Solving Systems of Two Linear Equations in Two Variables

A system of two linear equations in two variables is a set of equations that use the same variables (typically *x* and *y*). A solution to the system is an ordered pair that makes both equations true. One method for solving a system is by graphing. This method, however, is not always practical. Students may not have graph paper; or the solution may not consist of integers, making it difficult to identify the exact point of intersection on a graph. There are two methods for solving systems of equations algebraically: substitution and elimination. The method used will depend on the characteristics of the equations in the system.

Solving Systems of Equations with the Substitution Method

If one of the equations in a system has an isolated variable (*x*= or *y*=) or a variable that can be easily isolated, the substitution method can be used. Here's a sample system: $x + 3y = 7; 2x - 4y = 24$. The first equation can easily be solved for *x*. By subtracting 3*y* on both sides, the resulting equation is $x = 7 - 3y$. When one equation is solved for a variable, the expression that it is equal can be substituted into the other equation. For this example, $(7 - 3y)$ would be substituted for *x* into the second equation as follows: $2(7 - 3y) + 4y = 24$. Solving this equation results in $y = -5$. Once the value for one variable is known, this value should be substituted into either of the original equations to determine the value of the other variable. For the example, -5 would be substituted for *y* in either of the original equations. Substituting into the first equation results in $x + 3(-5) = 7$, and solving this equation yields $x = 22$. The solution to a system is an ordered pair, so the solution to the example is written as (22, 7). The solution can be checked by substituting it into both equations of the system to ensure it results in two true statements.

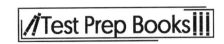

Solving Systems of Equations with the Elimination Method

The elimination method for solving a system of equations involves canceling out (or eliminating) one of the variables. This method is typically used when both equations of a system are written in standard form ($Ax + By = C$). An example is:

$$2x + 3y = 12; 5x - y = 13$$

To perform the elimination method, the equations in the system should be arranged vertically to be added together and then one or both of the equations should be multiplied so that one variable will be eliminated when the two are added. Opposites will cancel each other when added together. For example, $8x$ and $-8x$ will cancel each other when added. For the example above, writing the system vertically helps identify that the bottom equation should be multiplied by 3 to eliminate the variable y.

$$2x + 3y = 12 \quad \rightarrow \quad 2x + 3y = 12$$

$$3(5x - y = 13) \quad \rightarrow \quad 15x - 3y = 39$$

Adding the two equations together vertically results in $17x = 51$. Solving yields $x = 3$. Once the value for one variable is known, it can be substituted into either of the original equations to determine the value of the other variable. Once this is obtained, the solution can be written as an ordered pair (x, y) and checked in both equations of the system. In this example, the solution is (3, 2).

Systems of Equations with No Solution or an Infinite Number of Solutions

A system of equations can have one solution, no solution, or an infinite number of solutions. If, while solving a system algebraically, both variables cancel out, then the system has either no solution or has an infinite number of solutions. If the remaining constants result in a true statement (ex., $7 = 7$), then there is an infinite number of solutions. This would indicate coinciding lines. If the remaining constants result in a false statement, then there is no solution to the system. This would indicate parallel lines.

Linear Applications and Graphs

Linear Growth and Decay

As mentioned, a linear equation can be written in the form $y = mx + b$, where x represents the inputs, y represents the outputs, b represents the y-intercept for the graph, and m represents the slope of the line. The y-intercept is the value of y when $x = 0$ and can be thought of as the "starting point." The slope is the rate of change between the variables x and y. A positive slope represents growth, and a negative slope represents decay. Given a table of values for inputs (x) and outputs (y), a linear function would model the relationship if: x and y change at a constant rate per unit interval—for every two inputs a given distance apart, the distance between their corresponding outputs is constant. Here are some sample ordered pairs:

x	0	1	2	3
y	-7	-4	-1	2

For every 1 unit increase in x, y increases by 3 units. Therefore, the change is constant and thus represents linear growth.

Given a scenario involving growth or decay, determining if there is a constant rate of change between inputs (*x*) and outputs (*y*) will identify if a linear model is appropriate. A scenario involving distance and time might state that someone is traveling at a rate of 45 miles per hour. For every hour traveled (input), the distance traveled (output) increases by 45 miles. This is a constant rate of change.

Graphing Linear Equations

The simplest equations to graph are the equations whose graphs are lines, called **linear equations**. Every linear equation can be rewritten algebraically so that it looks like:

$$Ax + By = C$$

First, the ratio of the change in the *y* coordinate to the change in the *x* coordinate is constant for any two distinct points on the line. In any pair of points on a line, two points, (x_1, y_1) and (x_2, y_2)—

where $x_1 \neq x_2$—the ratio $\frac{y_2 - y_1}{x_2 - x_1}$ will always be the same, even if another pair of points is used.

This ratio, $\frac{y_2 - y_1}{x_2 - x_1}$, is called the *slope* of the line and is often denoted with the letter m. If the slope is **positive**, then the line goes upward when moving to the right. If the slope is **negative**, then it moves downward when moving to the right. If the slope is 0, then the line is **horizontal**, and the *y* coordinate is constant along the entire line. For lines where the *x* coordinate is constant along the entire line, the slope is not defined, and these lines are called **vertical** lines.

The *y* coordinate of the point where the line touches the *y*-axis is called the **y-intercept** of the line. It is often denoted by the letter b, used in the form of the linear equation $y = mx + b$. The *x* coordinate of the point where the line touches the *x*-axis is called the **x-intercept**. It is also called the *zero* of the line.

Suppose two lines have slopes m_1 and m_2. If the slopes are equal, $m_1 = m_2$, then the lines are **parallel**. Parallel lines never meet one another. If $m_1 = -\frac{1}{m_2}$, then the lines are called **perpendicular** or **orthogonal**. Their slopes can also be called opposite reciprocals of each other.

There are several convenient ways to write down linear equations. The common forms are listed here:

- **Standard Form**: $Ax + By = C$, where the slope is given by $\frac{-A}{B}$, and the *y*-intercept is given by $\frac{C}{B}$.

- **Slope-Intercept Form**: $y = mx + b$, where the slope is m, and the *y*-intercept is b.

- **Point-Slope Form**: $y - y_1 = m(x - x_1)$, where m is the slope, and (x_1, y_1) is any point on the line.

- **Two-Point Form**: $\frac{y - y_1}{x - x_1} = \frac{y_2 - y_1}{x_2 - x_1}$, where (x_1, y_1), and (x_2, y_2) are any two distinct points on the line.

- **Intercept Form**: $\frac{x}{x_1} + \frac{y}{y_1} = 1$, where x_1 is the *x*-intercept, and y_1 is the *y*-intercept.

Depending upon the given information, different forms of the linear equation can be easier to write down than others. When given two points, the two-point form is easy to write down. If the slope and a single point is known, the point-slope form is easiest to start with. In general, which form to start with depends upon the given information.

Graphing a Line in Slope-Intercept Form
When an equation is written in slope-intercept form, $y = mx + b$, m represents the slope of the line and b represents the y-intercept. The y-intercept is the value of y when $x = 0$ and the point at which the graph of the line crosses the y-axis. The slope is the rate of change between the variables, expressed as a fraction. The fraction expresses the change in y compared to the change in x. If the slope is an integer, it should be written as a fraction with a denominator of 1. For example, 5 would be written as 5/1.

To graph a line given an equation in slope-intercept form, the y-intercept should first be plotted. For example, to graph $y = -\frac{2}{3}x + 7$, the y-intercept of 7 would be plotted on the y-axis (vertical axis) at the point (0, 7). Next, the slope would be used to determine a second point for the line. Note that all that is necessary to graph a line is two points on that line. The slope will indicate how to get from one point on the line to another. The slope expresses vertical change (y) compared to horizontal change (x) and therefore is sometimes referred to as $\frac{rise}{run}$. The numerator indicates the change in the y value (move up for positive integers and move down for negative integers), and the denominator indicates the change in the x value. For the previous example, using the slope of $-\frac{2}{3}$, from the first point at the y-intercept, the second point should be found by counting down 2 and to the right 3. This point would be located at (3, 5).

Graphing a Line in Standard Form
When an equation is written in standard form, $Ax + By = C$, it is easy to identify the x- and y-intercepts for the graph of the line. Just as the y-intercept is the point at which the line intercepts the y-axis, the x-intercept is the point at which the line intercepts the x-axis. At the y-intercept, $x = 0$; and at the x-intercept, $y = 0$. Given an equation in standard form, $x = 0$ should be used to find the y-intercept. Likewise, $y = 0$ should be used to find the x-intercept. For example, to graph $3x + 2y = 6$, 0 for y results in $3x + 2(0) = 6$. Solving for y yields $x = 2$; therefore, an ordered pair for the line is (2, 0). Substituting 0 for x results in $3(0) + 2y = 6$. Solving for y yields $y = 3$; therefore, an ordered pair for

109

the line is (0, 3). The two ordered pairs (the *x*- and *y*-intercepts) can be plotted and a straight line through them can be constructed.

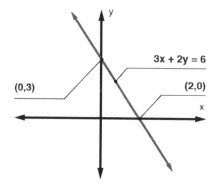

T - chart

x	y
0	3
2	0

Intercepts

x - intercept : (2,0)

y - intercept : (0,3)

Writing the Equation of a Line Given its Graph

Given the graph of a line, its equation can be written in two ways. If the *y*-intercept is easily identified (is an integer), it and another point can be used to determine the slope. When determining $\frac{change\ in\ y}{change\ in\ x}$ from one point to another on the graph, the distance for $\frac{rise}{run}$ is being figured. The equation should be written in slope-intercept form, $y = mx + b$, with *m* representing the slope and *b* representing the *y*-intercept.

The equation of a line can also be written by identifying two points on the graph of the line. To do so, the slope is calculated and then the values are substituted for the slope and either of the ordered pairs into the point-slope form of an equation.

Vertical, Horizontal, Parallel, and Perpendicular Lines

For a vertical line, the value of *x* remains constant (for all ordered pairs (*x*, *y*) on the line, the value of *x* is the same); therefore, the equations for all vertical lines are written in the form $x = number$. For example, a vertical line that crosses the *x*-axis at -2 would have an equation of $x = -2$. For a horizontal line, the value of *y* remains constant; therefore, the equations for all horizontal lines are written in the form $y = number$.

Parallel lines extend in the same exact direction without ever meeting. Their equations have the same slopes and different *y*-intercepts. For example, given a line with an equation of $y = -3x + 2$, a parallel line would have a slope of -3 and a *y*-intercept of any value other than 2. Perpendicular lines intersect to form a right angle. Their equations have slopes that are opposite reciprocal (the sign is changed and the fraction is flipped; for example, $-\frac{2}{3}$ and $\frac{3}{2}$) and *y*-intercepts that may or may not be the same. For example, given a line with an equation of $y = \frac{1}{2}x + 7$, a perpendicular line would have a slope of $-\frac{2}{1}$ and any value for its *y*-intercept.

110

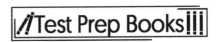

Graphing Solution Sets for Systems of Linear Inequalities in Two Variables

The solution set for a system of inequalities is the region of a graph consisting of ordered pairs that make both inequalities true. To graph the solution set, each linear inequality should first be graphed with appropriate shading. The region of the graph should be identified where the shading for the two inequalities overlaps. This region contains the solution set for the system.

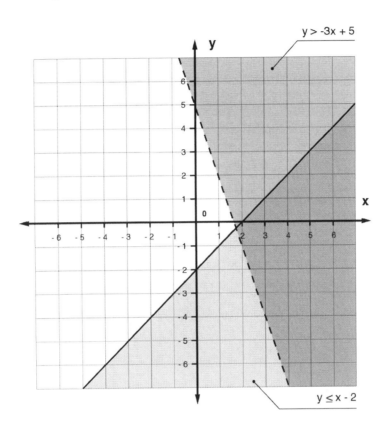

An ordered pair from the region of solutions can be selected to test in the system of inequalities.

Just as with manipulating linear inequalities in one variable, if dividing by a negative number in working with a linear inequality in two variables, the relationship reverses and the inequality sign should be flipped.

Factoring

Factors for polynomials are similar to factors for integers—they are numbers, variables, or polynomials that, when multiplied together, give a product equal to the polynomial in question. One polynomial is a factor of a second polynomial if the second polynomial can be obtained from the first by multiplying by a third polynomial.

$6x^6 + 13x^4 + 6x^2$ can be obtained by multiplying together $(3x^4 + 2x^2)(2x^2 + 3)$.

This means $2x^2 + 3$ and $3x^4 + 2x^2$ are factors of $6x^6 + 13x^4 + 6x^2$.

In general, finding the factors of a polynomial can be tricky. However, there are a few types of polynomials that can be factored in a straightforward way.

If a certain monomial is in each term of a polynomial, it can be factored out. There are several common forms polynomials take, which if you recognize, you can solve. The first example is a perfect square trinomial. To factor this polynomial, first expand the middle term of the expression:

$$x^2 + 2xy + y^2$$

$$x^2 + xy + xy + y^2$$

Factor out a common term in each half of the expression (in this case x from the left and y from the right):

$$x(x + y) + y(x + y)$$

Then the same can be done again, treating $(x + y)$ as the common factor:

$$(x + y)(x + y) = (x + y)^2$$

Therefore, the formula for this polynomial is:

$$x^2 + 2xy + y^2 = (x + y)^2$$

Next is another example of a perfect square trinomial. The process is the similar, but notice the difference in sign:

$$x^2 - 2xy + y^2$$

$$x^2 - xy - xy + y^2$$

Factor out the common term on each side:

$$x(x - y) - y(x - y)$$

Factoring out the common term again:

$$(x - y)(x - y) = (x - y)^2$$

Thus:

$$x^2 - 2xy + y^2 = (x - y)^2$$

The next is known as a difference of squares. This process is effectively the reverse of binomial multiplication:

$$x^2 - y^2$$

$$x^2 - xy + xy - y^2$$

$$x(x - y) + y(x - y)$$

$$(x + y)(x - y)$$

Therefore:

$$x^2 - y^2 = (x + y)(x - y)$$

The following two polynomials are known as the sum or difference of cubes. These are special polynomials that take the form of $x^3 + y^3$ or $x^3 - y^3$. The following formula factors the sum of cubes:

$$x^3 + y^3 = (x + y)(x^2 - xy + y^2)$$

Next is the difference of cubes, but note the change in sign. The formulas for both are similar, but the order of signs for factoring the sum or difference of cubes can be remembered by using the acronym SOAP, which stands for "same, opposite, always positive." The first sign is the same as the sign in the first expression, the second is opposite, and the third is always positive. The next formula factors the difference of cubes:

$$x^3 - y^3 = (x - y)(x^2 + xy + y^2)$$

The following two examples are expansions of cubed binomials. Similarly, these polynomials always follow a pattern:

$$x^3 + 3x^2y + 3xy^2 + y^3 = (x + y)^3$$

$$x^3 - 3x^2y + 3xy^2 - y^3 = (x - y)^3$$

These rules can be used in many combinations with one another. For example, the expression $3x^3 - 24$ has a common factor of 3, which becomes:

$$3(x^3 - 8)$$

A difference of cubes still remains which can then be factored out:

$$3(x - 2)(x^2 + 2x + 4)$$

There are no other terms to be pulled out, so this expression is completely factored.

When factoring polynomials, a good strategy is to multiply the factors to check the result. Let's try another example:

$$4x^3 + 16x^2$$

Both sides of the expression can be divided by 4, and both contain x^2, because $4x^3$ can be thought of as $4x^2(x)$, so the common term can simply be factored out:

$$4x^2(x + 4)$$

It sometimes can be necessary to rewrite the polynomial in some clever way before applying the above rules. Consider the problem of factoring $x^4 - 1$. This does not immediately look like any of the previous polynomials. However, it's possible to think of this polynomial as $x^4 - 1 = (x^2)^2 - (1^2)^2$, and now it can be treated as a difference of squares to simplify this:

$$(x^2)^2 - (1^2)^2$$

$$(x^2)^2 - x^2 1^2 + x^2 1^2 - (1^2)^2$$

$$x^2(x^2 - 1^2) + 1^2(x^2 - 1^2)$$

$$(x^2 + 1^2)(x^2 - 1^2)$$

$$(x^2 + 1)(x^2 - 1)$$

Factoring an Algebraic Expression

There are many different ways to write algebraic expressions and equations that are equivalent to each other. Converting expressions from standard form to factored form and vice versa are skills commonly used in advanced mathematics. Standard form of an expression arranges terms with variables powers in descending order (highest exponent to lowest and then constants). Factored form displays an expression as the product of its factors (what can be multiplied to produce the expression).

Converting Standard Form to Factored Form

To factor an expression, a greatest common factor needs to be factored out first. Then, if possible, the remaining expression needs to be factored into the product of binomials. A binomial is an expression with two terms.

Greatest Common Factor

The greatest common factor (GCF) of a monomial (one term) consists of the largest number that divides evenly into all coefficients (number part of a term), and if all terms contain the same variable, the variable with the lowest exponent. The GCF of $3x^4 - 9x^3 + 12x^2$ would be $3x^2$. To write the factored expression, every term needs to be divided by the GCF, then the product of the resulting quotient and the GCF (using parentheses to show multiplication) should be written. For the previous example, the factored expression would be:

$$3x^2(x^2 - 3x + 4)$$

Factoring Ax² + Bx + C When A = 1

To factor a quadratic expression in standard form when the value of a is equal to 1, the factors that multiply to equal the value of c should be found and then added to equal the value of b (the signs of b and c should be included). The factored form for the expression will be the product of binomials: $(x + factor1)(x + factor2)$. Here's a sample expression: $x^2 - 4x - 5$. The two factors that multiply to equal c(-5) and add together to equal b(-4) are -5 and 1. Therefore, the factored expression would be $(x - 5)(x + 1)$. Note $(x + 1)(x - 5)$ is equivalent.

Factoring a Difference of Squares

A difference of squares is a binomial expression where both terms are perfect squares (perfect square-perfect square). Perfect squares include 1, 4, 9, 16 . . . and x^2, x^4, x^6 . . .

The factored form of a difference of squares will be:

$$(\sqrt{term1} + \sqrt{term2})(\sqrt{term1} - \sqrt{term2})$$

For example:

$$x^2 - 4 = (x + 2)(x - 2)$$

And

$$25x^6 - 81 = (5x^3 + 9)(5x^3 - 9)$$

Factoring Ax² + Bx + C when A ≠ 1

To factor a quadratic expression in standard form when the value of a is not equal to 1, the factors that multiply to equal the value of $a \times c$ should be found and then added to equal the value of b. Next, the expression splitting the bx term should be rewritten using those factors. Instead of three terms, there will now be four. Then the first two terms should be factored using GCF, and a common binomial should be factored from the last two terms. The factored form will be: (common binomial) (2 terms out of binomials). In the sample expression $2 \times 2 + 11x + 12$, the value of $a \times c$ $(2x12) = 24$. Two factors that multiply to 24 and added together to yield $b(11)$ are 8 and 3. The bx term $(11x)$ can be rewritten by splitting it into the factors:

$$2 \times 2 + 8x + 3x + 12$$

A GCF from the first two terms can be factored as:

$$2x(x + 4) + 3x + 12$$

A common binomial from the last two terms can then be factored as:

$$2(x + 4) + 3(x + 4)$$

The factored form can be written as a product of binomials:

$$(x + 4)(2x + 3)$$

Converting Factored Form to Standard Form

To convert an expression from factored form to standard form, the factors are multiplied.

Quadratics

Quadratic Equations

A **quadratic equation** can be written in the form $y = ax^2 + bx + c$. The u-shaped graph of a quadratic function is called a **parabola**. The graph can either open up or open down (upside down u). The graph is symmetric about a vertical line, called the axis of symmetry. Corresponding points on the parabola are directly across from each other (same y-value) and are the same distance from the axis of symmetry (on either side). The axis of symmetry intersects the parabola at its vertex. The y-value of the vertex represents the minimum or maximum value of the function. If the graph opens up, the value of a in its equation is positive, and the vertex represents the minimum of the function. If the graph opens down, the value of *a* in its equation is negative, and the vertex represents the maximum of the function.

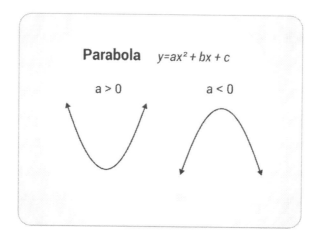

For a quadratic equation where the value of *a* is positive, as the inputs increase, the outputs increase until a certain value (maximum of the function) is reached. As inputs increase past the value that corresponds with the maximum output, the relationship reverses, and the outputs decrease. For a quadratic equation where *a* is negative, as the inputs increase, the outputs (1) decrease, (2) reach a maximum, and (3) then increase.

Consider a ball thrown straight up into the air. As time passes, the height of the ball increases until it reaches its maximum height. After reaching the maximum height, as time increases, the height of the ball decreases (it is falling toward the ground). This relationship can be expressed as a quadratic function where time is the input (*x*) and the height of the ball is the output (*y*).

To write a quadratic equation that models a real-life scenario, the following is needed: its vertex and any other ordered pair; or any three ordered pairs for the function. Given three ordered pairs, they should be substituted into the general form ($y = ax^2 + bx + c$) to create a system of three equations. For example, given the ordered pairs (2, 3), (3, 13), and (4, 29), it yields:

$$3 = a(2)2 + b(2) + c \rightarrow 4a + 2b + c = 3$$

$$13 = a(3)2 + b(3) + c \rightarrow 9a + 3b + c = 13$$

$$29 = a(4)2 + b(4) + c \rightarrow 16a + 24b + c = 29$$

The values for *a*, *b*, and *c* in the system can be found and substituted into the general form to write the equation of the function. In this case, the equation is $y = 3x^2 - 5x + 1$.

Given a curve of best-fit that models a quadratic relationship, the equation of the parabola can be written by identifying the vertex of the parabola and another point on the graph. The values for the vertex (h, k) and the point (x, y) should be substituted into the vertex form of a quadratic function, $y = a(x - h)^2 + k$, to determine the value of *a*. To write the equation of a quadratic model with a vertex of

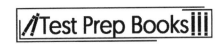

(4, 7) and containing the point (8, 3), the values for *h*, *k*, *x*, and *y* should be substituted into the vertex form of a quadratic equation, resulting in:

$$3 = a(8 - 4)^2 + 7$$

Solving for *a*, yields $a = -\frac{1}{4}$. Therefore, the equation can be written as:

$$y = -\frac{1}{4}(x - 4)^2 + 7$$

The vertex form can be manipulated in order to write the quadratic equation in standard form.

Solving a Quadratic Equation

A quadratic equation is one in which the highest exponent of the variable is 2. A quadratic equation can have two, one, or zero real solutions. Depending on its structure, a quadratic equation can be solved by (1) factoring, (2) taking square roots, or (3) using the quadratic formula.

Solving Quadratic Equations by Factoring
To solve a quadratic equation by factoring, the equation should first be manipulated to set the quadratic expression equal to zero. Next, the quadratic expression should be factored using the appropriate method(s). Then each factor should be set equal to zero. If two factors multiply to equal zero, then one or both factors must equal zero. Finally, each equation should be solved. Here's a sample: $x^2 - 10 = 3x - 6$. The expression should be set equal to zero: $x^2 - 3x - 4 = 0$. The expression should be factored:

$$(x - 4)(x + 1) = 0$$

Each factor should be set equal to zero: $x - 4 = 0; x + 1 = 0$. Solving yields $x = 4$ or $x = -1$.

Solving Quadratic Equations by Taking Square Roots
If a quadratic equation does not have a linear term (variable to the first power), it can be solved by taking square roots. This means x^2 needs to be isolated and then the square root of both sides of the equation should be isolated. There will be two solutions because square roots can be positive or negative. ($\sqrt{4}$ = 2 or -2 because $2 \times 2 = 4$ and $-2 \times -2 = 4$.) Here's a sample equation: $3x^2 - 12 = 0$. Isolating x^2 yields $x^2 = 4$. The square root of both sides is then solved: $x = 2$ or -2.

The Quadratic Formula
When a quadratic expression cannot be factored or is difficult to factor, the quadratic formula can be used to solve the equation. To do so, the equation must be in the form:

$$ax^2 + bx + c = 0$$

The quadratic formula is:

$$x = \frac{-b \pm \sqrt{b^2 - 4ac}}{2a}$$

(The $\pm$ symbol indicates that two calculations are necessary, one using + and one using −.) Here's a sample equation:

$$3x^2 - 2x = 3x + 2$$

First, the quadratic expression should be set equal to zero:

$$3x^2 - 5x - 2 = 0$$

Then the values are substituted for a(3), b(-5), and c(-2) into the formula:

$$x = \frac{-(-5) \pm \sqrt{(-5)^2 - 4(3)(-2)}}{2(3)}$$

Simplification yields:

$$x = \frac{5 \pm \sqrt{49}}{6} \rightarrow x = \frac{5 \pm 7}{6}$$

Calculating two values for x using + and − yields:

$$x = \frac{5 + 7}{6}; x = \frac{5 - 7}{6}$$

Simplification yields:

$$x = 2 \text{ or } -\frac{1}{3}.$$

Just as with any equation, solutions should be checked by substituting the value into the original equation.

Solving a System of One Linear Equation and One Quadratic Equation

A system of equations consists of two variables in two equations. A solution to the system is an ordered pair (x, y) that makes both equations true. When displayed graphically, a solution to a system is a point of intersection between the graphs of the equations. When a system consists of one linear equation and one quadratic equation, there may be one, two, or no solutions. If the line and parabola intersect at two points, there are two solutions to the system; if they intersect at one point, there is one solution; if they do not intersect, there is no solution.

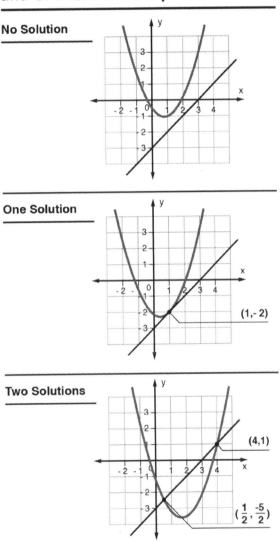

One method for solving a system of one linear equation and one quadratic equation is to graph both functions and identify point(s) of intersection. This, however, is not always practical. Graph paper may not be available, or the intersection points may not be easily identified. Solving the system algebraically involves using the substitution method. Consider the following system: $y = x^2 + 9x + 11; y = 2x - 1$.

The equivalent value of y should be substituted from the linear equation $(2x - 1)$ into the quadratic equation. The resulting equation is:

$$2x - 1 = x^2 + 9x + 11$$

Next, this quadratic equation should be solved using the appropriate method: factoring, taking square roots, or using the quadratic formula. Solving this quadratic equation by factoring results in $x = -4$ or $x = -3$. Next, the corresponding y-values should be found by substituting the x-values into the original linear equation: $y = 2(-4) - 1; y = 2(-3) - 1$. The solutions should be written as ordered pairs: (-4, -9) and (-3, -7). Finally, the possible solutions should be checked by substituting each into both of the original equations. In this case, both solutions "check out."

Functions

A **function** is defined as a relationship between inputs and outputs where there is only one output value for a given input. As an example, the following function is in function notation:

$$f(x) = 3x - 4$$

The $f(x)$ represents the output value for an input of x. If $x = 2$, the equation becomes:

$$f(2) = 3(2) - 4 = 6 - 4 = 2$$

The input of 2 yields an output of 2, forming the ordered pair $(2, 2)$. The following set of ordered pairs corresponds to the given function: $(2, 2), (0, -4), (-2, -10)$. The set of all possible inputs of a function is its **domain**, and all possible outputs is called the **range**. By definition, each member of the domain is paired with only one member of the range.

Functions can also be defined recursively. In this form, they are not defined explicitly in terms of variables. Instead, they are defined using previously-evaluated function outputs, starting with either $f(0)$ or $f(1)$. An example of a recursively-defined function is:

$$f(1) = 2, f(n) = 2f(n - 1) + 2n, n > 1$$

The domain of this function is the set of all integers.

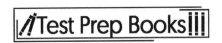

Creating Functions Using Function Notation

A **function** is defined as a relationship between inputs and outputs where there is only one output value for a given input. As an example, the following function is in function notation:

$$f(x) = 3x - 4$$

The $f(x)$ represents the output value for an input of x. If $x = 2$, the equation becomes:

$$f(2) = 3(2) - 4 = 6 - 4 = 2$$

The input of 2 yields an output of 2, forming the ordered pair $(2, 2)$. The following set of ordered pairs corresponds to the given function:

$$(2, 2), (0, -4), (-2, -10)$$

The set of all possible inputs of a function is its **domain**, and all possible outputs is called the **range**. By definition, each member of the domain is paired with only one member of the range.

Functions can also be defined recursively. In this form, they are not defined explicitly in terms of variables. Instead, they are defined using previously-evaluated function outputs, starting with either $f(0)$ or $f(1)$. An example of a recursively-defined function is:

$$f(1) = 2, f(n) = 2f(n - 1) + 2n, n > 1$$

The domain of this function is the set of all integers.

A function $f(x)$ is a mathematical object which takes one number, x, as an input and gives a number in return. The input is called the **independent variable**. If the variable is set equal to the output, as in $y = f(x)$, then this is called the **dependent variable**. To indicate the dependent value a function, y, gives for a specific independent variable, x, the notation y = $f(x)$ is used.

The **domain** of a function is the set of values that the independent variable is allowed to take. Unless otherwise specified, the domain is any value for which the function is well defined. The **range** of the function is the set of possible outputs for the function.

In many cases, a function can be defined by giving an equation. For instance, $f(x) = x^2$ indicates that given a value for x, the output of f is found by squaring x.

Not all equations in x and y can be written in the form $y = f(x)$. An equation can be written in such a form if it satisfies the **vertical line test**: no vertical line meets the graph of the equation at more than a single point. In this case, y is said to be a *function of x*. If a vertical line meets the graph in two places, then this equation cannot be written in the form $y = f(x)$.

The graph of a function $f(x)$ is the graph of the equation $y = f(x)$. Thus, it is the set of all pairs (x, y) where $y = f(x)$. In other words, it is all pairs $(x, f(x))$. The x-intercepts are called the **zeros** of the function. The y-intercept is given by $f(0)$.

If, for a given function f, the only way to get $f(a) = f(b)$ is for $a = b$, then f is *one-to-one*. Often, even if a function is not one-to-one on its entire domain, it is one-to-one by considering a restricted portion of the domain.

A function $f(x) = k$ for some number k is called a **constant function**. The graph of a constant function is a horizontal line.

The function $f(x) = x$ is called the **identity function**. The graph of the identity function is the diagonal line pointing to the upper right at 45 degrees, $y = x$.

A function is called **monotone** if it is either always increasing or always decreasing. For example, the functions $f(x) = 3x$ and $f(x) = -x^5$ are monotone.

An **even function** looks the same when flipped over the y-axis: $f(x) = f(-x)$. The following image shows a graphic representation of an even function.

Parabola

An **odd function** looks the same when flipped over the y-axis and then flipped over the x-axis: $f(x) = -f(-x)$. The following image shows an example of an odd function.

Domain and Range

The domain and range of a function can be found visually by its plot on the coordinate plane. In the function $f(x) = x^2 - 3$, for example, the domain is all real numbers because the parabola stretches as far left and as far right as it can go, with no restrictions. This means that any input value from the real number system will yield an answer in the real number system. For the range, the inequality $y \geq -3$ would be used to describe the possible output values because the parabola has a minimum at $y = -3$.

This means there will not be any real output values less than -3 because -3 is the lowest value it reaches on the y-axis.

These same answers for domain and range can be found by observing a table. The table below shows that from input values $x = -1$ to $x = 1$, the output results in a minimum of -3. On each side of $x = 0$, the numbers increase, showing that the range is all real numbers greater than or equal to -3.

x (domain/input)	y (range/output)
-2	1
-1	-2
0	-3
-1	-2
2	1

Function Behavior

Different types of functions behave in different ways. A function is defined to be increasing over a subset of its domain if for all $x_1 \geq x_2$ in that interval, $f(x_1) \geq f(x_2)$. Also, a function is decreasing over an interval if for all $x_1 \geq x_2$ in that interval, $f(x_1) \leq f(x_2)$. A point in which a function changes from increasing to decreasing can also be labeled as the **maximum value** of a function if it is the largest point the graph reaches on the y-axis. A point in which a function changes from decreasing to increasing can be labeled as the minimum value of a function if it is the smallest point the graph reaches on the y-axis. Maximum values are also known as **extreme values**. The graph of a continuous function does not have any breaks or jumps in the graph. This description is not true of all functions. A radical function, for example, $f(x) = \sqrt{x}$, has a restriction for the domain and range because there are no real negative inputs or outputs for this function. The domain can be stated as $x \geq 0$, and the range is $y \geq 0$.

A piecewise-defined function also has a different appearance on the graph. In the following function, there are three equations defined over different intervals. It is a function because there is only one y-value for each x-value, passing the Vertical Line Test. The domain is all real numbers less than or equal to 6. The range is all real numbers greater than zero. From left to right, the graph decreases to zero, then increases to almost 4, and then jumps to 6.

123

From input values greater than 2, the input decreases just below 8 to 4, and then stops.

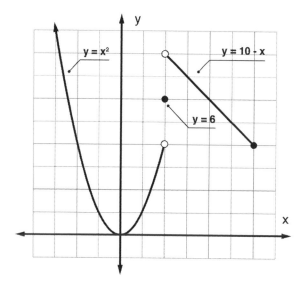

Logarithmic and exponential functions also have different behavior than other functions. These two types of functions are inverses of each other. The **inverse** of a function can be found by switching the place of x and y, and solving for y. When this is done for the exponential equation, $y = 2^x$, the function $y = \log_2 x$ is found. The general form of a **logarithmic function** is $y = \log_b x$, which says b raised to the y power equals x.

The thick black line on the graph above represents the logarithmic function:

$$y = \log_2 x$$

This curve passes through the point $(1, 0)$, just as all log functions do, because any value $b^0 = 1$. The graph of this logarithmic function starts very close to zero but does not touch the y-axis. The output value will never be zero by the definition of logarithms. The thinner gray line seen above represents the exponential function $y = 2^x$. The behavior of this function is opposite the logarithmic function because the graph of an inverse function is the graph of the original function flipped over the line $y = x$. The curve passes through the point $(0, 1)$ because any number raised to the zero power is one. This curve also gets very close to the x-axis but never touches it because an exponential expression never has an output of zero. The x-axis on this graph is called a horizontal asymptote. An **asymptote** is a line that represents a boundary for a function. It shows a value that the function will get close to, but never reach.

Functions can also be described as being even, odd, or neither. If $f(-x) = f(x)$, the function is even. For example, the function $f(x) = x^2 - 2$ is even. Plugging in $x = 2$ yields an output of $y = 2$. After changing the input to $x = -2$, the output is still $y = 2$. The output is the same for opposite inputs. Another way to observe an even function is by the symmetry of the graph. If the graph is symmetrical about the axis, then the function is even. If the graph is symmetric about the origin, then the function is odd. Algebraically, if $f(-x) = -f(x)$, the function is odd.

Also, a function can be described as **periodic** if it repeats itself in regular intervals. Common periodic functions are trigonometric functions. For example, $y = \sin x$ is a periodic function with period 2π because it repeats itself every 2π units along the x-axis.

Common Functions

Three common functions used to model different relationships between quantities are linear, quadratic, and exponential functions. **Linear functions** are the simplest of the three, and the independent variable x has an exponent of 1. Written in the most common form, $y = mx + b$, the coefficient of x indicates how fast the function grows at a constant rate, and the b-value denotes the starting point. A **quadratic** function has an exponent of 2 on the independent variable x. Standard form for this type of function is $y = ax^2 + bx + c$, and the graph is a parabola. These type functions grow at a changing rate. An **exponential** function has an independent variable in the exponent $y = ab^x$. The graph of these types of functions is described as **growth** or **decay**, based on whether the base, b, is greater than or less than 1. These functions are different from quadratic functions because the base stays constant. A common base is base e.

The following three functions model a linear, quadratic, and exponential function respectively: $y = 2x$, $y = x^2$, and $y = 2^x$. Their graphs are shown below. The first graph, modeling the linear function, shows that the growth is constant over each interval. With a horizontal change of 1, the vertical change is 2. It models a constant positive growth. The second graph shows the quadratic function, which is a curve that is symmetric across the y-axis. The growth is not constant, but the change is mirrored over the axis. The last graph models the exponential function, where the horizontal change of 1 yields a vertical change that increases more and more. The exponential graph gets very close to the x-axis, but never touches it, meaning there is an asymptote there. The y-value can never be zero because the base of 2 can never be raised to an input value that yields an output of zero.

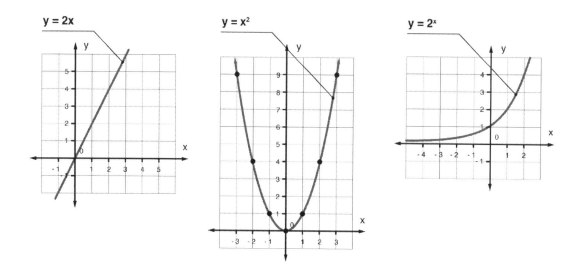

The three tables below show specific values for three types of functions. The third column in each table shows the change in the y-values for each interval. The first table shows a constant change of 2 for each equal interval, which matches the slope in the equation $y = 2x$. The second table shows an increasing change, but it also has a pattern. The increase is changing by 2 more each time, so the change is quadratic. The third table shows the change as factors of the base, 2. It shows a continuing pattern of factors of the base.

y = 2x		
x	y	$\triangle$y
1	2	
2	4	2
3	6	2
4	8	2
5	10	2

y = x²		
x	y	$\triangle$y
1	1	
2	4	3
3	9	5
4	16	7
5	25	9

y = 2ˣ		
x	y	$\triangle$y
1	2	
2	4	2
3	8	4
4	16	8
5	32	16

Given a table of values, the type of function can be determined by observing the change in y over equal intervals. For example, the tables below model two functions. The changes in interval for the x-values is 1 for both tables. For the first table, the y-values increase by 5 for each interval. Since the change is constant, the situation can be described as a linear function. The equation would be $y = 5x + 3$. For the second table, the change for y is 5, 20, 100, and 500, respectively. The increases are multiples of 5, meaning the situation can be modeled by an exponential function. The equation $y = 5^x + 3$ models this situation.

x	y
0	3
1	8
2	13
3	18
4	23

x	y
0	3
1	8
2	28
3	128
4	628

Quadratic equations can be used to model real-world area problems. For example, a farmer may have a rectangular field that he needs to sow with seed. The field has length $x + 8$ and width $2x$. The formula for area should be used: $A = lw$. Therefore:

$$A = (x + 8) \times 2x = 2x^2 + 16x$$

The possible values for the length and width can be shown in a table, with input x and output A. If the equation was graphed, the possible area values can be seen on the y-axis for given x-values.

Exponential growth and decay can be found in real-world situations. For example, if a piece of notebook paper is folded 25 times, the thickness of the paper can be found. To model this situation, a table can be used. The initial point is one-fold, which yields a thickness of 2 papers. For the second fold, the thickness is 4. Since the thickness doubles each time, the table below shows the thickness for the next few folds. Notice the thickness changes by the same factor each time. Since this change for a constant interval of folds is a factor of 2, the function is exponential. The equation for this is $y = 2^x$. For twenty-five folds, the thickness would be 33,554,432 papers.

x (folds)	y (paper thickness)
0	1
1	2
2	4
3	8
4	16
5	32

One exponential formula that is commonly used is the **interest formula**: $A = Pe^{rt}$. In this formula, interest is compounded continuously. A is the value of the investment after the time, t, in years. P is the initial amount of the investment, r is the interest rate, and e is the constant equal to approximately 2.718. Given an initial amount of $200 and a time of 3 years, if interest is compounded continuously at a rate of 6%, the total investment value can be found by plugging each value into the formula. The invested value at the end is $239.44. In more complex problems, the final investment may be given, and the rate may be the unknown. In this case, the formula becomes $239.44 = 200e^{r3}$. Solving for r requires isolating the exponential expression on one side by dividing by 200, yielding the equation

$1.20 = e^{r3}$. Taking the natural log of both sides results in $\ln(1.2) = r3$. Using a calculator to evaluate the logarithmic expression, $r = 0.06 = 6\%$.

When working with logarithms and exponential expressions, it is important to remember the relationship between the two. In general, the logarithmic form is $y = log_b x$ for an exponential form $b^y = x$. Logarithms and exponential functions are inverses of each other.

Interpreting and Building a Function Within a Context

Functions can be built out of the context of a situation. For example, the relationship between the money paid for a gym membership and the months that someone has been a member can be described through a function. If the one-time membership fee is $40 and the monthly fee is $30, then the function can be written $f(x) = 30x + 40$. The x-value represents the number of months the person has been part of the gym, while the output is the total money paid for the membership. The table below shows this relationship. It is a representation of the function because the initial cost is $40 and the cost increases each month by $30.

x (months)	y (money paid to gym)
0	40
1	70
2	100
3	130

Functions can also be built from existing functions. For example, a given function $f(x)$ can be transformed by adding a constant, multiplying by a constant, or changing the input value by a constant. The new function $g(x) = f(x) + k$ represents a vertical shift of the original function. In $f(x) = 3x - 2$, a vertical shift 4 units up would be:

$$g(x) = 3x - 2 + 4 = 3x + 2$$

Multiplying the function times a constant k represents a vertical stretch, based on whether the constant is greater than or less than 1. The function

$$g(x) = kf(x) = 4(3x - 2) = 12x - 8$$

represents a stretch. Changing the input x by a constant forms the function:

$$g(x) = f(x + k) = 3(x + 4) - 2$$

$$3x + 12 - 2 = 3x + 10$$

and this represents a horizontal shift to the left 4 units. If $(x - 4)$ was plugged into the function, it would represent a vertical shift.

A composition function can also be formed by plugging one function into another. In function notation, this is written:

$$(f \circ g)(x) = f(g(x))$$

For two functions $f(x) = x^2$ and $g(x) = x - 3$, the composition function becomes:

$$f(g(x)) = (x - 3)^2 = x^2 - 6x + 9$$

The composition of functions can also be used to verify if two functions are inverses of each other. Given the two functions $f(x) = 2x + 5$ and $g(x) = \frac{x-5}{2}$, the composition function can be found $(f \circ g)(x)$. Solving this equation yields:

$$f(g(x)) = 2\left(\frac{x - 5}{2}\right) + 5 = x - 5 + 5 = x$$

It also is true that $g(f(x)) = x$. Since the composition of these two functions gives a simplified answer of x, this verifies that $f(x)$ and $g(x)$ are inverse functions. The domain of $f(g(x))$ is the set of all x-values in the domain of $g(x)$ such that $g(x)$ is in the domain of $f(x)$. Basically, both $f(g(x))$ and $g(x)$ have to be defined.

To build an inverse of a function, $f(x)$ needs to be replaced with y, and the x and y values need to be switched. Then, the equation can be solved for y. For example, given the equation $y = e^{2x}$, the inverse can be found by rewriting the equation $x = e^{2y}$. The natural logarithm of both sides is taken down, and the exponent is brought down to form the equation:

$$\ln(x) = \ln(e)\, 2y$$

$\ln(e)=1$, which yields the equation $\ln(x) = 2y$. Dividing both sides by 2 yields the inverse equation

$$\frac{\ln(x)}{2} = y = f^{-1}(x)$$

The domain of an inverse function is the range of the original function, and the range of an inverse function is the domain of the original function. Therefore, an ordered pair (x, y) on either a graph or a table corresponding to $f(x)$ means that the ordered pair (y, x) exists on the graph of $f^{-1}(x)$. Basically, if $f(x) = y$, then $f^{-1}(y) = x$. For a function to have an inverse, it must be one-to-one. That means it must pass the **Horizontal Line Test**, and if any horizontal line passes through the graph of the function twice, a function is not one-to-one. The domain of a function that is not one-to-one can be restricted to an interval in which the function is one-to-one, to be able to define an inverse function.

Functions can also be formed from combinations of existing functions.

Given $f(x)$ and $g(x)$, the following can be built:

$$f + g$$
$$f - g$$
$$fg$$
$$\frac{f}{g}$$

The domains of $f + g, f - g,$ and fg are the intersection of the domains of f and g. The domain of $\frac{f}{g}$ is the same set, excluding those values that make $g(x) = 0$.

For example, if:

$$f(x) = 2x + 3$$

$$g(x) = x + 1$$

then

$$\frac{f}{g} = \frac{2x + 3}{x + 1}$$

Its domain is all real numbers except -1.

Radical and Rational Equations

Radical Equations

Equations with radicals containing numbers only as the radicand are solved the same way that an equation without a radical would be. For example:

$$3x + \sqrt{81} = 45$$

would be solved using the same steps as if solving $2x + 4 = 12$. Radical equations are those in which the variable is part of the radicand. For example, $\sqrt{5x + 1} - 6 = 0$ and $\sqrt{x - 3} + 5 = x$ would be considered radical equations.

To solve a radical equation, the radical should be isolated and both sides of the equation should be raised to the same power to cancel the radical. Raising both sides to the second power will cancel a square root, raising to the third power will cancel a cube root, etc. To solve $\sqrt{5x + 1} - 6 = 0$, the radical should be isolated first: $\sqrt{5x + 1} = 6$. Then both sides should be raised to the second power:

$$(\sqrt{5x + 1})^2 = (6)^2 \rightarrow 5x + 1 = 36$$

Lastly, the linear equation should be solved: $x = 7$.

If a radical equation contains a variable in the radicand and a variable outside of the radicand, it must be checked for extraneous solutions. An extraneous solution is one obtained by following the proper process for solving an equation but does not "check out" when substituted into the original equation. Here's a sample equation:

$$\sqrt{x-3} + 5 = x$$

Isolating the radical yields:

$$\sqrt{x-3} = x - 5$$

Next, both sides should be squared to cancel the radical:

$$(\sqrt{x-3})^2 = (x-5)^2 \rightarrow x - 3 = (x-5)(x-5)$$

The binomials should be multiplied:

$$x - 3 = x^2 - 10x + 25$$

The quadratic equation is then solved:

$$0 = x^2 - 11x + 28$$

$$0 = (x-7)(x-4)$$

$$x - 7 = 0; x - 4 = 0 \rightarrow x = 7 \text{ or } x = 4$$

To check for extraneous solutions, each answer can be substituted, one at a time, into the original equation. Substituting 7 for x, results in $7 = 7$. Therefore, 7 is a solution. Substituting 4 for x results in $6 = 4$. This is false; therefore, 4 is an extraneous solution.

Rational Expressions, Equations, and Functions

Rational Expressions

A **rational expression** is a fraction where the numerator and denominator are both polynomials. Some examples of rational expressions include the following: $\frac{4x^3 y^5}{3z^4}$, $\frac{4x^3 + 3x}{x^2}$, and $\frac{x^2 + 7x + 10}{x+2}$. Since these refer to expressions and not equations, they can be simplified but not solved. Exponents on the variables are restricted to whole numbers, which means roots and negative exponents are not included in rational expressions.

Rational expressions can be transformed by factoring. For example, the expression $\frac{x^2 - 5x + 6}{(x-3)}$ can be rewritten by factoring the numerator to obtain:

$$\frac{(x-3)(x-2)}{(x-3)}$$

Therefore, the common binomial $(x-3)$ can cancel so that the simplified expression is:

$$\frac{(x-2)}{1} = (x-2)$$

Additionally, other rational expressions can be rewritten to take on different forms. Some may be factorable in themselves, while others can be transformed through arithmetic operations. Rational

131

expressions are closed under addition, subtraction, multiplication, and division by a nonzero expression. **Closed** means that if any one of these operations is performed on a rational expression, the result will still be a rational expression. The set of all real numbers is another example of a set closed under all four operations.

Adding and subtracting rational expressions is based on the same concepts as adding and subtracting simple fractions. For both concepts, the denominators must be the same for the operation to take place. For example, here are two rational expressions:

$$\frac{x^3 - 4}{(x - 3)} + \frac{x + 8}{(x - 3)}$$

Since the denominators are both $(x - 3)$, the numerators can be combined by collecting like terms to form:

$$\frac{x^3 + x + 4}{(x - 3)}$$

If the denominators are different, they need to be made common (the same) by using the Least Common Denominator (LCD). Each denominator needs to be factored, and the LCD contains each factor that appears in any one denominator the greatest number of times it appears in any denominator. The original expressions need to be multiplied times a form of 1, which will turn each denominator into the LCD. This process is like adding fractions with unlike denominators. It is also important when working with rational expressions to define what value of the variable makes the denominator zero. For this particular value, the expression is undefined.

Multiplication of rational expressions is performed like multiplication of fractions. The numerators are multiplied; then, the denominators are multiplied. The final fraction is then simplified. The expressions are simplified by factoring and cancelling out common terms. In the following example, the numerator of the second expression can be factored first to simplify the expression before multiplying:

$$\frac{x^2}{(x - 4)} \times \frac{x^2 - x - 12}{2}$$

$$\frac{x^2}{(x - 4)} \times \frac{(x - 4)(x + 3)}{2}$$

The $(x - 4)$ on the top and bottom cancel out:

$$\frac{x^2}{1} \times \frac{(x + 3)}{2}$$

Then multiplication is performed, resulting in:

$$\frac{x^3 + 3x^2}{2}$$

Dividing rational expressions is similar to the division of fractions, where division turns into multiplying by a reciprocal. So the following expression can be rewritten as a multiplication problem:

$$\frac{x^2 - 3x + 7}{x - 4} \div \frac{x^2 - 5x + 3}{x - 4}$$

$$\frac{x^2 - 3x + 7}{x - 4} \times \frac{x - 4}{x^2 - 5x + 3}$$

The $x - 4$ cancels out, leaving:

$$\frac{x^2 - 3x + 7}{x^2 - 5x + 3}$$

The final answers should always be completely simplified. If a function is composed of a rational expression, the zeros of the graph can be found from setting the polynomial in the numerator as equal to zero and solving. The values that make the denominator equal to zero will either exist on the graph as a hole or a vertical asymptote.

The following problem is an example of using rational expressions:

Reggie wants to lay sod in his rectangular backyard. The length of the yard is given by the expression $4x + 2$ and the width is unknown. The area of the yard is $20x + 10$. Reggie needs to find the width of the yard. Knowing that the area of a rectangle is length multiplied by width, an expression can be written to find the width: $\frac{20x+10}{4x+2}$, area divided by length. Simplifying this expression by factoring out 10 on the top and 2 on the bottom leads to this expression: $\frac{10(2x+1)}{2(2x+1)}$. Cancelling out the $2x + 1$ results in $\frac{10}{2} = 5$. The width of the yard is found to be 5 by simplifying the rational expression.

Rational Equations

A **rational equation** can be as simple as an equation with a ratio of polynomials, $\frac{p(x)}{q(x)}$, set equal to a value, where $p(x)$ and $q(x)$ are both polynomials. A rational equation has an equal sign, which is different from expressions. This leads to solutions, or numbers that make the equation true.

It is possible to solve rational equations by trying to get all of the x terms out of the denominator and then isolating them on one side of the equation. For example, to solve the equation $\frac{3x+2}{2x+3} = 4$, both sides get multiplied by $(2x + 3)$. This will cancel on the left side to yield: $3x + 2 = 4(2x + 3)$, then $3x + 2 = 8x + 12$. Now, subtract $8x$ from both sides, which yields $-5x + 2 = 12$. Subtracting 2 from both sides results in $-5x = 10$. Finally, both sides get divided by -5 to obtain $x = -2$.

Sometimes, when solving rational equations, it can be easier to try to simplify the rational expression by factoring the numerator and denominator first, then cancelling out common factors. For example, to solve $\frac{2x^2-8x+6}{x^2-3x+2} = 1$, the first step is to factor:

$$2x^2 - 8x + 6 = 2(x^2 - 4x + 3) = 2(x - 1)(x - 3)$$

Then, factor $x^2 - 3x + 2$ into $(x - 1)(x - 2)$. This turns the original equation into:

$$\frac{2(x - 1)(x - 3)}{(x - 1)(x - 2)} = 1$$

The common factor of $(x - 1)$ can be canceled, leaving:

$$\frac{2(x - 3)}{x - 2} = 1$$

Now the same method used in the previous example can be followed. Multiplying both sides by $x - 1$ and performing the multiplication on the left yields $2x - 6 = x - 2$, which can be simplified to $x = 4$.

Rational Functions

A **rational function** is similar to an equation, but it includes two variables. In general, a rational function is in the form: $f(x) = \frac{p(x)}{q(x)}$, where $p(x)$ and $q(x)$ are polynomials. Rational functions are defined everywhere except where the denominator is equal to zero. When the denominator is equal to zero, this indicates either a hole in the graph or an asymptote. An example of a function with an asymptote is shown below.

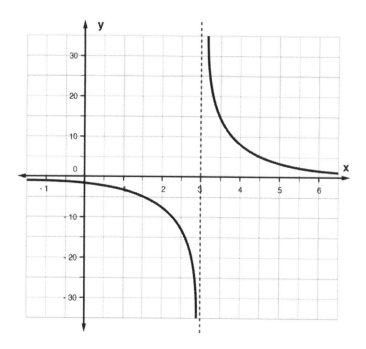

Polynomial Equations

Polynomials

An expression of the form ax^n, where n is a non-negative integer, is called a **monomial** because it contains one term. A sum of monomials is called a **polynomial**. For example, $-4x^3 + x$ is a polynomial, while $5x^7$ is a monomial. A function equal to a polynomial is called a **polynomial function**.

The monomials in a polynomial are also called the **terms** of the polynomial.

The constants that precede the variables are called **coefficients**.

The highest value of the exponent of x in a polynomial is called the **degree** of the polynomial. So, $-4x^3 + x$ has a degree of 3, while $-2x^5 + x^3 + 4x + 1$ has a degree of 5. When multiplying polynomials, the degree of the result will be the sum of the degrees of the two polynomials being multiplied.

To add polynomials, add the coefficients of like powers of x. For example:

$$(-2x^5 + x^3 + 4x + 1) + (-4x^3 + x)$$

$$-2x^5 + (1 - 4)x^3 + (4 + 1)x + 1$$

$$-2x^5 - 3x^3 + 5x + 1$$

Likewise, subtraction of polynomials is performed by subtracting coefficients of like powers of x. So:

$$(-2x^5 + x^3 + 4x + 1) - (-4x^3 + x)$$

$$-2x^5 + (1 + 4)x^3 + (4 - 1)x + 1$$

$$-2x^5 + 5x^3 + 3x + 1$$

To multiply two polynomials, multiply each term of the first polynomial by each term of the second polynomial and add the results. For example:

$$(4x^2 + x)(-x^3 + x)$$

$$4x^2(-x^3) + 4x^2(x) + x(-x^3) + x(x)$$

$$-4x^5 + 4x^3 - x^4 + x^2$$

In the case where each polynomial has two terms, like in this example, some students find it helpful to remember this as multiplying the First terms, then the Outer terms, then the Inner terms, and finally the Last terms, with the mnemonic **FOIL**. For longer polynomials, the multiplication process is the same, but there will be, of course, more terms, and there is no common mnemonic to remember each combination.

The process of **factoring** a polynomial means to write the polynomial as a product of other (generally simpler) polynomials. Here is an example:

$$x^2 - 4x + 3 = (x - 1)(x - 3)$$

If a certain monomial divides every term of the polynomial, factor it out of each term, for example:

$$4x^3 + 16x^2 = 4x^2(x + 4).$$

$$x^2 + 2xy + y^2 = (x + y)^2 \text{ or } x^2 - 2xy + y^2 = (x - y)^2$$

$$x^2 - y^2 = (x + y)(x - y)$$

$$x^3 + y^3 = (x + y)(x^2 - xy + y^2)$$

$$x^3 - y^3 = (x - y)(x^2 + xy + y^2)$$

$$x^3 + 3x^2y + 3xy^2 + y^3 = (x + y)^3 \text{ and } x^3 - 3x^2y + 3xy^2 - y^3 = (x - y)^3$$

It sometimes can be necessary to rewrite the polynomial in some clever way before applying the above rules. Consider the problem of factoring $x^4 - 1$. This does not immediately look like any of the cases for which there are rules. However, it's possible to think of this polynomial as $x^4 - 1 = (x^2)^2 - (1^2)^2$, and now apply the third rule in the above list to simplify this:

$$(x^2)^2 - (1^2)^2 = (x^2 + 1^2)(x^2 - 1^2) = (x^2 + 1)(x^2 - 1)$$

Zeros of Polynomials

Finding the zeros of polynomial equations are the points at which the graph of the equation crosses the x-axis. Factors can be used to find the zeros of a polynomial equation or function. The degree of the equation shows the number of possible zeros. If the highest exponent on the independent variable is 4, then the degree is 4, and the number of possible zeros is 4. If there are complex solutions, the number of roots is less than the degree.

Given the equation $y = x^2 + 7x + 6$, y can be set equal to zero, and the polynomial can be factored. The equation turns into $0 = (x + 1)(x + 6)$, where $x = -1$ and $x = -6$ are the zeros. Since this is a quadratic equation, the shape of the graph will be a parabola. Knowing that zeros represent the points where the parabola crosses the x-axis, the maximum or minimum point is the only other piece needed to sketch a rough graph of the function. By looking at the function in standard form, the coefficient of x is positive; therefore, the parabola opens *up*. Using the zeros and the minimum, the following rough sketch of the graph can be constructed:

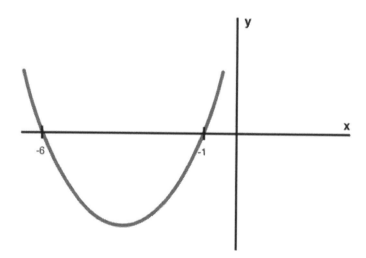

Polynomial Identities

Difference of squares refers to a binomial composed of the difference of two squares. For example, $a^2 - b^2$ is a difference of squares. It can be written $(a)^2 - (b)^2$, and it can be factored into $(a - b)(a + b)$. Recognizing the difference of squares allows the expression to be rewritten easily because of the form it takes. For some expressions, factoring consists of more than one step. When factoring, it's important to always check to make sure that the result cannot be factored further. If it can, then the

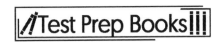

expression should be split further. If it cannot be, the factoring step is complete, and the expression is completely factored.

A sum and difference of cubes is another way to factor a polynomial expression. When the polynomial takes the form of addition or subtraction of two terms that can be written as a cube, a formula is given. The following graphic shows the factorization of a difference of cubes:

$$a^3 - b^3 = (a - b)(a^2 + ab + b^2)$$

same sign

opposite sign

always +

This form of factoring can be useful in finding the zeros of a function of degree 3. For example, when solving $x^3 - 27 = 0$, this rule needs to be used. $x^3 - 27$ is first written as the difference two cubes, $(x)^3 - (3)^3$ and then factored into;

$$(x - 3)(x^2 + 3x + 9)$$

This expression may not be factored any further. Each factor is then set equal to zero. Therefore, one solution is found to be $x = 3$, and the other two solutions must be found using the quadratic formula. A sum of squares would have a similar process. The formula for factoring a sum of squares is

$$a^3 + b^3 = (a + b)(a^2 - ab + b^2)$$

The opposite of factoring is multiplying. Multiplying a square of a binomial involves the following rules:

$$(a + b)^2 = a^2 + 2ab + b^2$$

and

$$(a - b)^2 = a^2 - 2ab + b^2$$

The binomial theorem for expansion can be used when the exponent on a binomial is larger than 2, and the multiplication would take a long time. The binomial theorem is given as:

$$(a + b)^n = \sum_{k=0}^{n} \binom{n}{k} a^{n-k} b^k \qquad \binom{n}{k} = \frac{n!}{k!(n-k)!}$$

where

The **Remainder Theorem** can be helpful when evaluating polynomial functions $P(x)$ for a given value of x. A polynomial can be divided by $(x - a)$, if there is a remainder of 0. This also means that $P(a) = 0$ and $(x - a)$ is a factor of $P(x)$. In a similar sense, if P is evaluated at any other number b, $P(b)$ is equal to the remainder of dividing $P(x)$ by $(x - b)$.

Exponential and Logarithmic Equations

Exponential Functions or Equations

An **exponential function** is a function of the form $f(x) = b^x$, where b is a positive real number other than 1. In such a function, b is called the **base**. An exponential equation takes the same form, but is typically written as $y = b^x$.

The **domain** of an exponential function is all real numbers, and the *range* is all positive real numbers. There will always be a horizontal asymptote of $y = 0$ on one side. If b is greater than 1, then the graph will be increasing moving to the right. If b is less than 1, then the graph will be decreasing moving to the right. Exponential functions are one-to-one. The basic exponential function graph will go through the point (0,1).

Example: Solve $5^{x+1} = 25$.

Get the x out of the exponent by rewriting the equation $5^{x+1} = 5^2$ so that both sides have a base of 5.

Since the bases are the same, the exponents must be equal to each other.

This leaves $x + 1 = 2$ or $x = 1$.

To check the answer, the x-value of 1 can be substituted back into the original equation.

Logarithmic Functions and Equations

A **logarithmic function** is an inverse for an exponential function. The inverse of the base b exponential function is written as $\log_b(x)$, and is called the *base b* **logarithm**. The domain of a logarithm is all positive real numbers. It has the properties that $\log_b(b^x) = x$. For positive real values of x, $b^{\log_b(x)} = x$.

When there is no chance of confusion, the parentheses are sometimes skipped for logarithmic functions: $\log_b(x)$ may be written as $\log_b x$. For the special number e, the base e logarithm is called the **natural logarithm** and is written as $\ln x$. Logarithms are one-to-one.

When working with logarithmic functions, it is important to remember the following properties. Each one can be derived from the definition of the logarithm as the inverse to an exponential function:

$$\log_b 1 = 0$$

$$\log_b b = 1$$

$$\log_b b^p = p$$

$$\log_b MN = \log_b M + \log_b N$$

$$\log_b \frac{M}{N} = \log_b M - \log_b N$$

$$\log_b M^p = p \log_b M$$

When solving equations involving exponentials and logarithms, the following fact should be used:

If f is a one-to-one function, $a = b$ is equivalent to $f(a) = f(b)$.

Using this, together with the fact that logarithms and exponentials are inverses, allows manipulations of the equations to isolate the variable.

Example: Solve $4 = \ln(x - 4)$.

Using the definition of a logarithm, the equation can be changed to $e^4 = e^{\ln(x-4)}$.

The functions on the right side cancel with a result of $e^4 = x - 4$.

This then gives $x = 4 + e^4$.

Trigonometry

Trigonometric Functions

Trigonometric functions are built out of two basic functions, the sine and cosine, written as $\sin \theta$ and $\cos \theta$ respectively. Note that similar to logarithms, it is customary to drop the parentheses as long as the result is not confusing.

The sine and cosine are defined using the **unit circle**. If θ is the angle going counterclockwise around the origin from the x-axis, then the point on the unit circle in that direction will have the coordinates $(\cos \theta, \sin \theta)$.

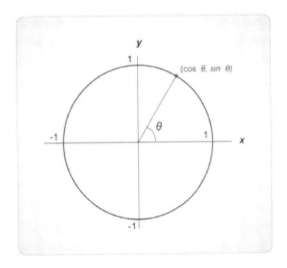

Since the angle returns to the start every 2π radians (or 360 degrees), the graph of these functions will be **periodic**, with period 2π. This means that the graph repeats itself as one moves along the x-axis because $\sin \theta = \sin(\theta + 2\pi)$. Cosine is works similarly.

From the unit circle definition, the **sine** function starts at 0 when $\theta = 0$. It grows to 1 as θ grows to $\pi/2$, and then back to 0 at $\theta = \pi$. Then it decreases to -1 as θ grows to $3\pi/2$, and back up to 0 at $\theta = 2\pi$.

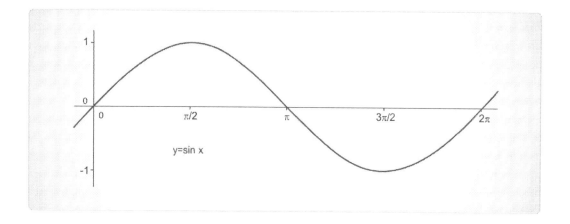

The graph of the **cosine** is similar. The cosine will start at 1, decreasing to 0 at $\pi/2$ and continuing to decrease to -1 at $\theta = \pi$. Then, it grows to 0 as θ grows to $3\pi/2$ and back up to 1 at $\theta = 2\pi$.

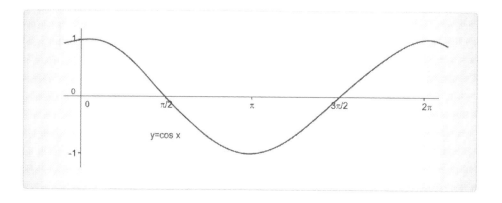

Another trigonometric function, which is frequently used, is the **tangent** function. This is defined as the following equation: $\tan \theta = \frac{\sin \theta}{\cos \theta}$.

The tangent function is a period of π rather than 2π because the sine and cosine functions have the same absolute values after a change in the angle of π, but flip their signs. Since the tangent is a ratio of the two functions, the changes in signs cancel.

The tangent function will be zero when the sine is zero, and it will have a vertical asymptote whenever cosine is zero. The following graph shows the tangent function:

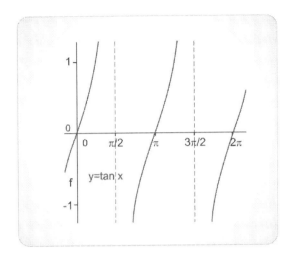

Three other trigonometric functions are sometimes useful. These are the *reciprocal trigonometric* **functions**, so named because they are just the reciprocals of sine, cosine, and tangent. They are the **cosecant**, defined as $\csc \theta = \frac{1}{\sin \theta}$, the **secant**, $\sec \theta = \frac{1}{\cos \theta}$, and the **cotangent**, $\cot \theta = \frac{1}{\tan \theta}$. Note that from the definition of tangent, $\cot \theta = \frac{\cos \theta}{\sin \theta}$.

In addition, there are three identities that relate the trigonometric functions to one another:

$$\cos \theta = \sin(\frac{\pi}{2} - \theta)$$

$$\csc \theta = \sec \left(\frac{\pi}{2} - \theta\right)$$

$$\cot \theta = \tan(\frac{\pi}{2} - \theta)$$

Here is a list of commonly-needed values for trigonometric functions, given in radians, for the first quadrant:

Table for trigonometric functions

$\sin 0 = 0$	$\cos 0 = 1$	$\tan 0 = 0$
$\sin \dfrac{\pi}{6} = \dfrac{1}{2}$	$\cos \dfrac{\pi}{6} = \dfrac{\sqrt{3}}{2}$	$\tan \dfrac{\pi}{6} = \dfrac{\sqrt{3}}{3}$
$\sin \dfrac{\pi}{4} = \dfrac{\sqrt{2}}{2}$	$\cos \dfrac{\pi}{4} = \dfrac{\sqrt{2}}{2}$	$\tan \dfrac{\pi}{4} = 1$
$\sin \dfrac{\pi}{3} = \dfrac{\sqrt{3}}{2}$	$\cos \dfrac{\pi}{3} = \dfrac{1}{2}$	$\tan \dfrac{\pi}{3} = \sqrt{3}$
$\sin \dfrac{\pi}{2} = 1$	$\cos \dfrac{\pi}{2} = 0$	$\tan \dfrac{\pi}{2} = undefined$
$\csc 0 = undefined$	$\sec 0 = 1$	$\cot 0 = undefined$
$\csc \dfrac{\pi}{6} = 2$	$\sec \dfrac{\pi}{6} = \dfrac{2\sqrt{3}}{3}$	$\cot \dfrac{\pi}{6} = \sqrt{3}$
$\csc \dfrac{\pi}{4} = \sqrt{2}$	$\sec \dfrac{\pi}{4} = \sqrt{2}$	$\cot \dfrac{\pi}{4} = 1$
$\csc \dfrac{\pi}{3} = \dfrac{2\sqrt{3}}{3}$	$\sec \dfrac{\pi}{3} = 2$	$\cot \dfrac{\pi}{3} = \dfrac{\sqrt{3}}{3}$
$\csc \dfrac{\pi}{2} = 1$	$\sec \dfrac{\pi}{2} = undefined$	$\cot \dfrac{\pi}{2} = 0$

To find the trigonometric values in other quadrants, complementary angles can be used. The **complementary angle** is the smallest angle between the x-axis and the given angle.

Once the complementary angle is known, the following rule is used:

For an angle θ with complementary angle x, the absolute value of a trigonometry function evaluated at θ is the same as the absolute value when evaluated at x.

The correct sign is used based on the functions sine and cosine are given by the x and y coordinates on the unit circle.

- Sine will be positive in quadrants I and II and negative in quadrants III and IV.

- Cosine will be positive in quadrants I and IV, and negative in II and III.

142

- Tangent will be positive in I and III, and negative in II and IV.

- The signs of the reciprocal functions will be the same as the sign of the function of which they are a reciprocal.

Example: Find $\sin\frac{3\pi}{4}$.

First, the complementary angle must be found.

This angle is in the II quadrant, and the angle between it and the x-axis is $\frac{\pi}{4}$.

Now, $\sin\frac{\pi}{4} = \frac{\sqrt{2}}{2}$.

Since this is in the II quadrant, sine takes on positive values (the y coordinate is positive in the II quadrant).

Therefore, $\sin\frac{3\pi}{4} = \frac{\sqrt{2}}{2}$.

In addition to the six trigonometric functions defined above, there are inverses for these functions. However, since the trigonometric functions are not one-to-one, one can only construct inverses for them on a restricted domain.

Usually, the domain chosen will be $[0, \pi)$ for cosine and $(-\frac{\pi}{2}, \frac{\pi}{2}]$ for sine. The inverse for tangent can use either of these domains. The inverse functions for the trigonometric functions are also called *arc functions*. In addition to being written with a -1 in the exponent to denote that the function is an inverse, they will sometimes be written with an "a" or "arc" in front of the function name, so $\cos^{-1}\theta = a\cos\theta = \arccos\theta$.

When solving equations that involve trigonometric functions, there are often multiple solutions. For example, $2\sin\theta = \sqrt{2}$ can be simplified to $\sin\theta = \frac{\sqrt{2}}{2}$. This has solutions $\theta = \frac{\pi}{4}, \frac{3\pi}{4}$, but in addition, because of the periodicity, any integer multiple of 2π can also be added to these solutions to find another solution.

The full set of solutions is $\theta = \frac{\pi}{4} + 2\pi k, \frac{3\pi}{4} + 2\pi k$ for all integer values of k. It is very important to remember to find all possible solutions when dealing with equations that involve trigonometric functions.

The name *trigonometric* comes from the fact that these functions play an important role in the geometry of triangles, particularly right triangles.

Consider the right triangle shown in this figure:

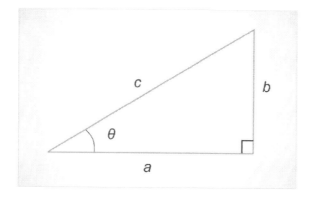

The following hold true:

- $c \sin \theta = b$

- $c \cos \theta = a$

- $\tan \theta = \frac{b}{a}$

- $b \csc \theta = c$

- $a \sec \theta = c$

- $\cot \theta = \frac{a}{b}$

Remember also the angles of a triangle must add up to π radians (180 degrees).

Solving Trigonometric Functions

Solving trigonometric functions can be done with a knowledge of the unit circle and the trigonometric identities. It requires the use of opposite operations combined with trigonometric ratios for special triangles. For example, the problem may require solving the equation:

$$2 \cos^2 x - \sqrt{3} \cos x = 0$$

for the values of x between 0 and 180 degrees. The first step is to factor out the $\cos x$ term, resulting in $\cos x (2 \cos x - \sqrt{3}) = 0$. By the factoring method of solving, each factor can be set equal to zero: $\cos x = 0$ and $(2 \cos x - \sqrt{3}) = 0$. The second equation can be solved to yield the following equation: $\cos x = \frac{\sqrt{3}}{2}$. Now that the value of x is found, the trigonometric ratios can be used to find the solutions of $x = 30$ and 90 degrees.

Solving trigonometric functions requires the use of algebra to isolate the variable and a knowledge of trigonometric ratios to find the value of the variable. The unit circle can be used to find answers for special triangles. Beyond those triangles, a calculator can be used to solve for variables within the trigonometric functions.

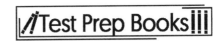

Using Graphing Calculators to Solve Trigonometric Problems

In addition to algebraic techniques, problems involving trigonometric functions can be solved using graphing calculators. For example, given an equation, both sides of the equals sign first need to be graphed as separate equations in the same window on the calculator. The point(s) of intersection are then found by zooming into an appropriate window. The points of intersection are the solutions. For example, consider $\cos x = \frac{1}{2}$. Solving this equation involves graphing both $y = \cos x$ and $y = \frac{1}{2}$ in the same window. If the calculator is set in Radians mode, the screen zoomed in from $[0, 2\pi]$ on the x-axis would look like:

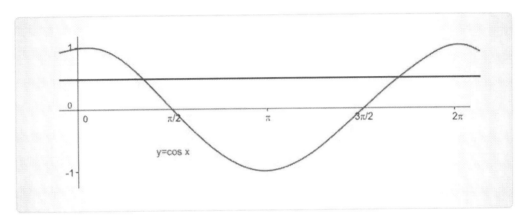

The trace function on the calculator allows the user to zoom into the point of intersection to obtain the two solutions from $[0, 2\pi]$. The periodic nature of the function must then be taken into consideration to obtain the entire solution set, which contains an infinite number of solutions.

Practice Questions

1. What is the product of the following expression?

$$(4x - 8)(5x^2 + x + 6)$$

    a. $20x^3 - 36x^2 + 16x - 48$
    b. $6x^3 - 41x^2 + 12x + 15$
    c. $204 + 11x^2 - 37x - 12$
    d. $2x^3 - 11x^2 - 32x + 20$

2. What is the y-intercept for $y = x^2 + 3x - 4$?
    a. $y = 1$
    b. $y = -4$
    c. $y = 3$
    d. $y = 4$

3. What is the value of b in the equation: $5b - 4 = 2b + 17$?
    a. 13
    b. 24
    c. 7
    d. 21

4. If $-3(x + 4) \geq x + 8$, what is the value of x?
    a. $x = 4$
    b. $x \geq 2$
    c. $x \geq -5$
    d. $x \leq -5$

5. Karen gets paid a weekly salary and a commission for every sale that she makes. The table below shows the number of sales and her pay for different weeks.

Sales	2	7	4	8
Pay	$380	$580	$460	$620

Which of the following equations represents Karen's weekly pay?
    a. $y = 90x + 200$
    b. $y = 90x - 200$
    c. $y = 40x + 300$
    d. $y = 40x - 300$

Answer Explanations

1. A: Finding the product means distributing one polynomial over the other so that each term in the first is multiplied by each term in the second. Then, like terms can be collected. Multiplying the factors yields the expression:

$$20x^3 + 4x^2 + 24x - 40x^2 - 8x - 48$$

Collecting like terms means adding the x^2 terms and adding the x terms. The final answer after simplifying the expression is:

$$20x^3 - 36x^2 + 16x - 48$$

By making this change in position over time into a rate, the speed becomes ten meters in two seconds or five meters in one second.

2. B: The y-intercept of an equation is found where the x-value is zero. Plugging zero into the equation for x, the first two terms cancel out, leaving -4.

3. C: To solve for the value of b, both sides of the equation need to be equalized.

Start by cancelling out the lower value of -4 by adding 4 to both sides:

$$5b - 4 = 2b + 17$$

$$5b = 2b + 21$$

The variable b is the same on each side, so subtract the lower 2b from each side:

$$5b = 2b + 21$$

$$3b = 21$$

Then divide both sides by 3 to get the value of b:

$$\frac{3b}{3} = \frac{21}{3}$$

$$b = 7$$

4. D: $x \leq -5$. When solving a linear equation or inequality:

Distribution is performed if necessary:

$$-3(x + 4) \rightarrow -3x - 12 \geq x + 8$$

This means that any like terms on the same side of the equation/inequality are combined.

The equation/inequality is manipulated to get the variable on one side. In this case, subtracting x from both sides produces:

$$-4x - 12 \geq 8$$

The variable is isolated using inverse operations to undo addition/subtraction. Adding 12 to both sides produces $-4x \geq 20$.

The variable is isolated using inverse operations to undo multiplication/division. Remember if dividing by a negative number, the relationship of the inequality reverses, so the sign is flipped. In this case, dividing by -4 on both sides produces $x \leq -5$.

5. C: $y = 40x + 300$

In this scenario, the variables are the number of sales and Karen's weekly pay. The weekly pay depends on the number of sales. Therefore, weekly pay is the dependent variable (y), and the number of sales is the independent variable (x). Each pair of values from the table can be written as an ordered pair (x, y): (2, 380), (7, 580), (4, 460), (8, 620).

The ordered pairs can be substituted into the equations to see which creates true statements (both sides equal) for each pair. Even if one ordered pair produces equal values for a given equation, the other three ordered pairs must be checked. The only equation which is true for all four ordered pairs is

$$y = 40x + 300$$

$$380 = 40(2) + 300 \rightarrow 380 = 380$$

$$580 = 40(7) + 300 \rightarrow 580 = 580$$

$$460 = 40(4) + 300 \rightarrow 460 = 460$$

$$620 = 40(8) + 300 \rightarrow 620 = 620$$

Practice Test #1

Arithmetic

1. 3.4+2.35+4=
    a. 5.35
    b. 9.2
    c. 9.75
    d. 10.25

2. $5.88 \times 3.2 =$
    a. 18.816
    b. 16.44
    c. 20.352
    d. 17

3. $\frac{3}{25} =$
    a. 0.15
    b. 0.1
    c. 0.9
    d. 0.12

4. Which of the following is largest?
    a. 0.45
    b. 0.096
    c. 0.3
    d. 0.313

5. Which of the following is NOT a way to write 40 percent of N?
    a. $(0.4)N$
    b. $\frac{2}{5}N$
    c. $40N$
    d. $\frac{4N}{10}$

6. Which is closest to 17.8×9.9?
    a. 140
    b. 180
    c. 200
    d. 350

7. A student gets an 85% on a test with 20 questions. How many answers did the student solve correctly?
    a. 15
    b. 16
    c. 17
    d. 18

8. Four people split a bill. The first person pays for $\frac{1}{5}$, the second person pays for $\frac{1}{4}$, and the third person pays for $\frac{1}{3}$. What fraction of the bill does the fourth person pay?

    a. $\frac{13}{60}$

    b. $\frac{47}{60}$

    c. $\frac{1}{4}$

    d. $\frac{4}{15}$

9. 6 is 30% of what number?

    a. 18

    b. 20

    c. 24

    d. 26

10. $3\frac{2}{3} - 1\frac{4}{5} =$

    a. $1\frac{13}{15}$

    b. $\frac{14}{15}$

    c. $2\frac{2}{3}$

    d. $\frac{4}{5}$

11. What is $\frac{420}{98}$ rounded to the nearest integer?

    a. 4

    b. 3

    c. 5

    d. 6

12. $4\frac{1}{3} + 3\frac{3}{4} =$

    a. $6\frac{5}{12}$

    b. $8\frac{1}{12}$

    c. $8\frac{2}{3}$

    d. $7\frac{7}{12}$

13. Five of six numbers have a sum of 25. The average of all six numbers is 6. What is the sixth number?

    a. 8

    b. 10

    c. 11

    d. 12

14. $52.3 \times 10^{-3} =$
 a. 0.00523
 b. 0.0523
 c. 0.523
 d. 523

15. If $\frac{5}{2} \div \frac{1}{3} = n$, then n is between:
 a. 5 and 7
 b. 7 and 9
 c. 9 and 11
 d. 3 and 5

16. A closet is filled with red, blue, and green shirts. If $\frac{1}{3}$ of the shirts are green and $\frac{2}{5}$ are red, what fraction of the shirts are blue?
 a. $\frac{4}{15}$

 b. $\frac{1}{5}$

 c. $\frac{7}{15}$

 d. $\frac{1}{2}$

17. Shawna buys $2\frac{1}{2}$ gallons of paint. If she uses $\frac{1}{3}$ of it on the first day, how much does she have left?
 a. $1\frac{5}{6}$ gallons

 b. $1\frac{1}{2}$ gallons

 c. $1\frac{2}{3}$ gallons

 d. 2 gallons

18. Mom's car drove 72 miles in 90 minutes. How fast did she drive in feet per second?
 a. 0.8 feet per second
 b. 48.9 feet per second
 c. 0.009 feet per second
 d. 70.4 feet per second

19. What is the simplified form of the expression $1.2 \times 10^{12} \div 3.0 \times 10^{8}$?
 a. 0.4×10^{4}
 b. 4.0×10^{4}
 c. 4.0×10^{3}
 d. 3.6×10^{20}

20. You measure the width of your door to be 36 inches. The true width of the door is 35.75 inches. What is the relative error in your measurement?

    a. 0.7%

    b. 0.007%

    c. 0.99%

    d. 0.1%

Quantitative Reasoning, Algebra, and Statistics

1. Which of the following is the result of simplifying the expression: $\frac{4a^{-1}b^3}{a^4b^{-2}} \times \frac{3a}{b}$?

    a. $12a^3b^5$

    b. $12\frac{b^4}{a^4}$

    c. $\frac{12}{a^4}$

    d. $7\frac{b^4}{a}$

2. Write the expression for three times the sum of twice a number and one minus 6.

    a. $2x + 1 - 6$

    b. $3x + 1 - 6$

    c. $3(x + 1) - 6$

    d. $3(2x + 1) - 6$

3. On Monday, Robert mopped the floor in 4 hours. On Tuesday, he did it in 3 hours. If on Monday, his average rate of mopping was p sq. ft. per hour, what was his average rate on Tuesday?

    a. $\frac{4}{3}p$ sq. ft. per hour

    b. $\frac{3}{4}p$ sq. ft. per hour

    c. $\frac{5}{4}p$ sq. ft. per hour

    d. $p + 1$ sq. ft. per hour

4. The phone bill is calculated each month using the equation $c = 50g + 75$. The cost of the phone bill per month is represented by c, and g represents the gigabytes of data used that month. What is the value and interpretation of the slope of this equation?

    a. 75 dollars per day

    b. 75 gigabytes per day

    c. 50 dollars per day

    d. 50 dollars per gigabyte

5. For the following similar triangles, what are the values of x and y (rounded to one decimal place)?

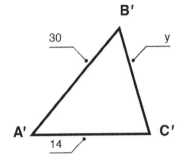

a. $x = 16.5, y = 25.1$
b. $x = 19.5, y = 24.1$
c. $x = 17.1, y = 26.3$
d. $x = 26.3, y = 17.1$

6. Solve for x, if $x^2 - 2x - 8 = 0$.
 a. $2 \pm \frac{\sqrt{30}}{2}$
 b. $2 \pm 4\sqrt{2}$
 c. 1 ± 3
 d. $4 \pm \sqrt{2}$

7. Which of the following is a factor of both $x^2 + 4x + 4$ and $x^2 - x - 6$?
 a. $x - 3$
 b. $x + 2$
 c. $x - 2$
 d. $x + 3$

8. What is the solution to the following system of equations?

$$x^2 - 2x + y = 8$$

$$x - y = -2$$

a. $(-2, 3)$
b. There is no solution.
c. $(-2, 0) (1, 3)$
d. $(-2, 0) (3, 5)$

9. What are the center and radius of a circle with equation $4x^2 + 4y^2 - 16x - 24y + 51 = 0$?
 a. Center $(3, 2)$ and radius $1/2$
 b. Center $(2, 3)$ and radius $1/2$
 c. Center $(3, 2)$ and radius $1/4$
 d. Center $(2, 3)$ and radius $1/4$

10. What's the midpoint of a line segment with endpoints $(-1, 2)$ and $(3, -6)$?
   a. $(1, 2)$
   b. $(1, 0)$
   c. $(-1, 2)$
   d. $(1, -2)$

11. A sample data set contains the following values: 1, 3, 5, 7. What's the standard deviation of the set?
   a. 2.58
   b. 4
   c. 6.23
   d. 1.1

12. A ball is drawn at random from a ball pit containing 8 red balls, 7 yellow balls, 6 green balls, and 5 purple balls. What's the probability that the ball drawn is yellow?
   a. $\frac{1}{26}$
   b. $\frac{19}{26}$
   c. $\frac{7}{26}$
   d. 1

13. Two cards are drawn from a shuffled deck of 52 cards. What's the probability that both cards are Kings if the first card isn't replaced after it's drawn?
   a. $\frac{1}{169}$
   b. $\frac{1}{221}$
   c. $\frac{1}{13}$
   d. $\frac{4}{13}$

14. What's the probability of rolling a 6 at least once in two rolls of a die?
   a. $\frac{1}{3}$
   b. $\frac{1}{36}$
   c. $\frac{1}{6}$
   d. $\frac{11}{36}$

15. A line goes through the point (-4, 0) and the point (0,2). What is the slope of the line?
   a. 2
   b. 4
   c. $\frac{3}{2}$
   d. $\frac{1}{2}$

16. Six people apply to work for Janice's company, but she only needs four workers. How many different groups of four employees can Janice choose?
- a. 6
- b. 10
- c. 15
- d. 36

17. $2x(3x + 1) - 5(3x + 1) =$
- a. $10x(3x + 1)$

- b. $10x^2(3x + 1)$

- c. $(2x - 5)(3x + 1)$

- d. $(2x + 1)(3x - 5)$

18. For which real numbers x is $-3x^2 + x - 8 > 0$?
- a. All real numbers x

- b. $-2\sqrt{\frac{2}{3}} < x < 2\sqrt{\frac{2}{3}}$

- c. $1 - 2\sqrt{\frac{2}{3}} < x < 1 + 2\sqrt{\frac{2}{3}}$

- d. For no real numbers x

19. A root of $x^2 - 2x - 2$ is
- a. $1 + \sqrt{3}$
- b. $1 + 2\sqrt{2}$
- c. $2 + 2\sqrt{3}$
- d. $2 - 2\sqrt{3}$

20. In the xy-plane, the graph of $y = x^2 + 2$ and the circle with center $(0,1)$ and radius 1 have how many points of intersection?
- a. 0
- b. 1
- c. 2
- d. 3

Advanced Algebra and Functions

1. What is the value of $x^2 - 2xy + 2y^2$ when $x = 2, y = 3$?
- a. 8
- b. 10
- c. 12
- d. 14

2. $(2x - 4y)^2 =$
    a. $4x^2 - 16xy + 16y^2$
    b. $4x^2 - 8xy + 16y^2$
    c. $4x^2 - 16xy - 16y^2$
    d. $2x^2 - 8xy + 8y^2$

3. If $x > 3$, then $\frac{x^2 - 6x + 9}{x^2 - x - 6} =$
    a. $\frac{x+2}{x-3}$

    b. $\frac{x-2}{x-3}$

    c. $\frac{x-3}{x+3}$

    d. $\frac{x-3}{x+2}$

4. The square and circle have the same center. The circle has a radius of r. What is the area of the shaded region?

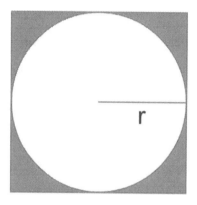

    a. $r^2 - \pi r^2$
    b. $4r^2 - 2\pi r$
    c. $(4 - \pi)r^2$
    d. $(\pi - 1)r^2$

5. If $4x - 3 = 5$, then $x =$
    a. 1
    b. 2
    c. 3
    d. 4

6. What are the zeros of the function: $f(x) = x^3 + 4x^2 + 4x$?
    a. -2
    b. $0, -2$
    c. 2
    d. $0, 2$

7. If $g(x) = x^3 - 3x^2 - 2x + 6$ and $f(x) = 2$, then what is $g(f(x))$?

    a. −26

    b. 6

    c. $2x^3 - 6x^2 - 4x + 12$

    d. −2

8. What is the function that forms an equivalent graph to $y = \cos(x)$?

    a. $y = \tan(x)$

    b. $y = \csc(x)$

    c. $y = \sin(x + \frac{\pi}{2})$

    d. $y = \sin(x - \frac{\pi}{2})$

9. What is the solution for the equation $\tan(x) + 1 = 0$, where $0 \le x < 2\pi$?

    a. $x = \frac{3\pi}{4}, \frac{5\pi}{4}$

    b. $x = \frac{3\pi}{4}, \frac{\pi}{4}$

    c. $x = \frac{5\pi}{4}, \frac{7\pi}{4}$

    d. $x = \frac{3\pi}{4}, \frac{7\pi}{4}$

10. Which of the following inequalities is equivalent to $3 - \frac{1}{2}x \ge 2$?

    a. $x \ge 2$

    b. $x \le 2$

    c. $x \ge 1$

    d. $x \le 1$

11. For which of the following are $x = 4$ and $x = -4$ solutions?

    a. $x^2 + 16 = 0$

    b. $x^2 + 4x - 4 = 0$

    c. $x^2 - 2x - 2 = 0$

    d. $x^2 - 16 = 0$

12. If x is not zero, then $\frac{3}{x} + \frac{5u}{2x} - \frac{u}{4} =$

    a. $\frac{12 + 10u - ux}{4x}$

    b. $\frac{3 + 5u - ux}{x}$

    c. $\frac{12x + 10u + ux}{4x}$

    d. $\frac{12 + 10u - u}{4x}$

13. Which of the following shows the correct result of simplifying the following expression?

$$(7n + 3n^3 + 3) + (8n + 5n^3 + 2n^4)$$

a. $9n^4 + 15n - 2$
b. $2n^4 + 5n^3 + 15n - 2$
c. $9n^4 + 8n^3 + 15n$
d. $2n^4 + 8n^3 + 15n + 3$

14. How could the following equation be factored to find the zeros?

$$y = x^3 - 3x^2 - 4x$$

a. $0 = x^2(x - 4), x = 0, 4$
b. $0 = 3x(x + 1)(x + 4), x = 0, -1, -4$
c. $0 = x(x + 1)(x + 6), x = 0, -1, -6$
d. $0 = x(x + 1)(x - 4), x = 0, -1, 4$

15. What is the simplified quotient of $\frac{5x^3}{3x^2y} \div \frac{25}{3y^9}$?

a. $\frac{125x}{9y^{10}}$
b. $\frac{x}{5y^8}$
c. $\frac{5}{xy^8}$
d. $\frac{xy^8}{5}$

16. What type of function is modeled by the values in the following table?

x	$f(x)$
1	2
2	4
3	8
4	16
5	32

a. Linear
b. Exponential
c. Quadratic
d. Cubic

17. What is the simplified form of the expression $tan\theta\ cos\theta$?
a. $sin\theta$
b. 1
c. $csc\theta$
d. $\frac{1}{sec\theta}$

18. Which equation is not a function?
    a. $y = |x|$
    b. $y = \sqrt{x}$
    c. $x = 3$
    d. $y = 4$

19. How could the following function be rewritten to identify the zeros?

$$y = 3x^3 + 3x^2 - 18x$$

    a. $y = 3x(x + 3)(x - 2)$
    b. $y = x(x - 2)(x + 3)$
    c. $y = 3x(x - 3)(x + 2)$
    d. $y = (x + 3)(x - 2)$

20. If $x^2 + x - 3 = 0$, then $\left(x - \frac{1}{2}\right)^2 =$
    a. $\dfrac{11}{2}$

    b. $\dfrac{11}{4}$

    c. 11

    d. $\dfrac{121}{4}$

Answer Explanations #1

Arithmetic

1. C: The decimal points are lined up, with zeroes put in as needed. Then, the numbers are added just like integers:

$$
\begin{array}{r}
3.40 \\
2.35 \\
+4.00 \\
\hline
9.75
\end{array}
$$

2. A: This problem can be multiplied as 588×32, except at the end, the decimal point needs to be moved three places to the left. Performing the multiplication will give 18,816, and moving the decimal place over three places results in 18.816.

3. D: The fraction is converted so that the denominator is 100 by multiplying the numerator and denominator by 4, to get $\frac{3}{25} = \frac{12}{100}$. Dividing a number by 100 just moves the decimal point two places to the left, with a result of 0.12.

4. A: Figure out which is largest by looking at the first non-zero digits. Choice *B*'s first non-zero digit is in the hundredths place. The other three all have non-zero digits in the tenths place, so it must be *A*, *C*, or *D*. Of these, *A* has the largest first non-zero digit.

5. C: $40N$ would be 4000% of *N*. It's possible to check that each of the others is actually 40% of *N*.

6. B: Instead of multiplying these out, the product can be estimated by using $18 \times 10 = 180$. The error here should be lower than 15, since it is rounded to the nearest integer, and the numbers add to something less than 30.

7. C: 85% of a number means multiplying that number by 0.85. So, $0.85 \times 20 = \frac{85}{100} \times \frac{20}{1}$, which can be simplified to:

$$
\frac{17}{20} \times \frac{20}{1} = 17
$$

8. A: To find the fraction of the bill that the first three people pay, the fractions need to be added, which means finding the common denominator. The common denominator will be 60.

$$
\frac{1}{5} + \frac{1}{4} + \frac{1}{3}
$$

$$
\frac{12}{60} + \frac{15}{60} + \frac{20}{60} = \frac{47}{60}
$$

The remainder of the bill is:

$$
1 - \frac{47}{60} = \frac{60}{60} - \frac{47}{60} = \frac{13}{60}
$$

9. B: 30% is 3/10. The number itself must be 10/3 of 6, or:

$$\frac{10}{3} \times 6 = 10 \times 2 = 20$$

10. A: First, convert the mixed numbers to improper fractions: $\frac{11}{3} - \frac{9}{5}$. Then, use 15 as a common denominator:

$$\frac{11}{3} - \frac{9}{5}$$

$$\frac{55}{15} - \frac{27}{15} = \frac{28}{15} = 1\frac{13}{15}$$

11. A: Dividing by 98 can be approximated by dividing by 100, which would mean shifting the decimal point of the numerator to the left by 2. The result is 4.2 and rounds to 4.

12. B:

$$4\frac{1}{3} + 3\frac{3}{4} = 4 + 3 + \frac{1}{3} + \frac{3}{4}$$

$$7 + \frac{1}{3} + \frac{3}{4}$$

Adding the fractions gives:

$$\frac{1}{3} + \frac{3}{4} = \frac{4}{12} + \frac{9}{12}$$

$$\frac{13}{12} = 1 + \frac{1}{12}$$

Thus:

$$7 + \frac{1}{3} + \frac{3}{4}$$

$$7 + 1 + \frac{1}{12} = 8\frac{1}{12}$$

13. C: The average is calculated by adding all six numbers, then dividing by 6. The first five numbers have a sum of 25. If the total divided by 6 is equal to 6, then the total itself must be 36. The sixth number must be 36 − 25 = 11.

14. B: Multiplying by 10^{-3} means moving the decimal point three places to the left, putting in zeroes as necessary.

15. B: $\frac{5}{2} \div \frac{1}{3} = \frac{5}{2} \times \frac{3}{1} = \frac{15}{2} = 7.5$

16. A: The total fraction taken up by green and red shirts will be:

$$\frac{1}{3} + \frac{2}{5} = \frac{5}{15} + \frac{6}{15} = \frac{11}{15}$$

The remaining fraction is:

$$1 - \frac{11}{15} = \frac{15}{15} - \frac{11}{15} = \frac{4}{15}$$

17. C: If she has used 1/3 of the paint, she has 2/3 remaining. $2\frac{1}{2}$ gallons are the same as $\frac{5}{2}$ gallons. The calculation is:

$$\frac{2}{3} \times \frac{5}{2} = \frac{5}{3} = 1\frac{2}{3} \text{ gallons}$$

18. D: This problem can be solved by using unit conversions. The initial units are miles per minute. The final units need to be feet per second. Converting miles to feet uses the equivalence statement $1 \text{ } mile = 5,280 \text{ } feet$. Converting minutes to seconds uses the equivalence statement $1 \text{ } minute = 60 \text{ } seconds$. Setting up the ratios to convert the units is shown in the following equation:

$$\frac{72 \text{ } miles}{90 \text{ } minutes} \times \frac{1 \text{ } minute}{60 \text{ } seconds} \times \frac{5280 \text{ } feet}{1 \text{ } mile} = 70.4 \text{ } feet \text{ } per \text{ } second$$

The initial units cancel out, and the new, desired units are left.

19. C: Scientific notation division can be solved by grouping the first terms together and grouping the tens together. The first terms can be divided, and the tens terms can be simplified using the rules for exponents. The initial expression becomes 0.4×10^4. This is not in scientific notation because the first number is not between 1 and 10. Shifting the decimal and subtracting one from the exponent, the answer becomes 4.0×10^3.

20. A: The relative error can be found by finding the absolute error and making it a percent of the true value. The absolute error is $36 - 35.75 = 0.25$. This error is then divided by 35.75—the true value—to find 0.7%.

Quantitative Reasoning, Algebra, and Statistics

1. B: To simplify the given equation, the first step is to make all exponents positive by moving them to the opposite place in the fraction. This expression becomes $\frac{4b^3b^2}{a^1a^4} \times \frac{3a}{b}$. Then the rules for exponents can be used to simplify. Multiplying the same bases means the exponents can be added. Dividing the same bases means the exponents are subtracted. Thus, after multiplying the exponents in the first fraction the equation becomes $\frac{4b^5}{a^5} \times \frac{3a}{b}$. Therefore, we can first multiply to get $\frac{12ab^5}{a^5b}$. Then, dividing yields $12\frac{b^4}{a^4}$.

2. D: The expression is three times the sum of twice a number and 1, which is $3(2x + 1)$. Then, 6 is subtracted from this expression.

3. A: Robert accomplished his task on Tuesday in $\frac{3}{4}$ the time compared to Monday. He must have worked $\frac{4}{3}$ as fast.

4. D: The slope from this equation is 50, and it is interpreted as the cost per gigabyte used. Since the g-value represents number of gigabytes and the equation is set equal to the cost in dollars, the slope relates these two values. For every gigabyte used on the phone, the bill goes up 50 dollars.

5. C: Because the triangles are similar, the lengths of the corresponding sides are proportional. Therefore:

$$\frac{30+x}{30} = \frac{22}{14} = \frac{y+15}{y}$$

This results in the equation $14(30+x) = 22 \times 30$ which, when solved, gives $x = 17.1$. The proportion also results in the equation $14(y+15) = 22y$ which, when solved, gives $y = 26.3$.

6. C: The numbers needed are those that add to -2 and multiply to -8. The difference between 2 and 4 is 2. Their product is 8, and -4 and 2 will work. Therefore:

$$x^2 - 2x - 8 = (x-4)(x+2)$$

The latter has roots 4 and -2 or 1 ± 3.

7. B: To factor $x^2 + 4x + 4$, the numbers needed are those that add to 4 and multiply to 4. Therefore, both numbers must be 2, and the expression factors to:

$$x^2 + 4x + 4 = (x+2)^2$$

Similarly, the second expression factors to $x^2 - x - 6 = (x-3)(x+2)$, so that they have $x+2$ in common.

8. D: This system of equations involves one quadratic function and one linear function, as seen from the degree of each equation. One way to solve this is through substitution. Solving for y in the second equation yields $y = x+2$. Plugging this equation in for the y of the quadratic equation yields;

$$x^2 - 2x + x + 2 = 8$$

Simplifying the equation, it becomes $x^2 - x + 2 = 8$. Setting this equal to zero and factoring, it becomes:

$$x^2 - x - 6 = 0 = (x-3)(x+2)$$

Solving these two factors for x gives the zeros $x = 3, -2$. To find the y-value for the point, each number can be plugged in to either original equation. Solving each one for y yields the points $(3, 5)$ and $(-2, 0)$.

9. B: The technique of completing the square must be used to change

$$4x^2 + 4y^2 - 16x - 24y + 51 = 0$$

into the standard equation of a circle. First, the constant must be moved to the right-hand side of the equals sign, and each term must be divided by the coefficient of the x^2 term (which is 4). The x and y terms must be grouped together to obtain:

$$x^2 - 4x + y^2 - 6y = -\frac{51}{4}$$

Then, the process of completing the square must be completed for each variable. This gives:

$$(x^2 - 4x + 4) + (y^2 - 6y + 9) = -\frac{51}{4} + 4 + 9$$

The equation can be written as $(x - 2)^2 + (y - 3)^2 = \frac{1}{4}$. Therefore, the center of the circle is $(2, 3)$ and the radius is $\sqrt{1/4} = 1/2$.

10. D: The midpoint formula should be used to get the average of both points.

$$M = \left(\frac{x_1 + x_2}{2}, \frac{y_1 + y_2}{2}\right) = \left(\frac{-1 + 3}{2}, \frac{2 + (-6)}{2}\right) = (1, -2)$$

11. A: First, the sample mean must be calculated. $\bar{x} = \frac{1}{4}(1 + 3 + 5 + 7) = 4$. The sample standard deviation of the data set is:

$$s = \sqrt{\frac{\sum(x - \bar{x})^2}{n - 1}}$$

and $n = 4$ represents the number of data points.

Therefore, the sample standard deviation is:

$$s = \sqrt{\frac{1}{3}[(1 - 4)^2 + (3 - 4)^2 + (5 - 4)^2 + (7 - 4)^2]}$$

$$s = \sqrt{\frac{1}{3}(9 + 1 + 1 + 9)} = 2.58$$

12. C: The sample space is made up of $8 + 7 + 6 + 5 = 26$ balls. The probability of pulling each individual ball is $1/26$. Since there are 7 yellow balls, the probability of pulling a yellow ball is $7/26$.

13. B: For the first card drawn, the probability of a King being pulled is $\frac{4}{52}$. Since this card isn't replaced, if a King is drawn first the probability of a King being drawn second is $\frac{3}{51}$. The probability of a King being drawn in both the first and second draw is the product of the two probabilities: $\frac{4}{52} \times \frac{3}{51} = \frac{12}{2652}$. This fraction, when divided by 12, equals $\frac{1}{221}$.

14. D: The addition rule is necessary to determine the probability because a 6 can be rolled on either roll of the die. The rule used is:

$$P(A \text{ or } B) = P(A) + P(B) - P(A \text{ and } B)$$

The probability of a 6 being individually rolled is $\frac{1}{6}$ and the probability of a 6 being rolled twice is:

$$\frac{1}{6} \times \frac{1}{6} = \frac{1}{36}$$

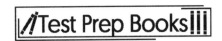

Therefore, the probability that a 6 is rolled at least once is:

$$\frac{1}{6} + \frac{1}{6} - \frac{1}{36} = \frac{11}{36}$$

15. D: The slope is given by the change in *y* divided by the change in *x*. The change in *y* is 2-0 = 2, and the change in *x* is:

$$0 - (-4) = 4$$

The slope is $\frac{2}{4} = \frac{1}{2}$.

16. C: Janice will be choosing 4 employees out of a set of 6 applicants, so this will be given by the choice function. The following equation shows the choice function worked out:

$$\binom{6}{4} = \frac{6!}{4!\,(6-4)!} = \frac{6!}{4!\,(2)!}$$

$$\frac{6 \cdot 5 \cdot 4 \cdot 3 \cdot 2 \cdot 1}{4 \cdot 3 \cdot 2 \cdot 1 \cdot 2 \cdot 1} = \frac{6 \cdot 5}{2} = 15$$

17. C: The $(3x + 1)$ can be factored to get $(2x - 5)(3x + 1)$.

18. D: Because the coefficient of x^2 is negative, this function has a graph that is a parabola that opens downward. Therefore, it will be greater than 0 between its real roots, if it has any. Checking the discriminant, the result is:

$$1^2 - 4(-3)(-8) = 1 - 96 = -95$$

Since the discriminant is negative, this equation has no real solutions. Since this has no real roots, it must be always positive or always negative. Its graph opens downward, so it has at least some negative values. That means it is always negative. Thus, it is greater than zero for no real numbers.

19. A: Check each value, but it is easiest to use the quadratic formula, which gives

$$x = \frac{2 \pm \sqrt{(-2)^2 - 4(1)(-2)}}{2} = 1 \pm \frac{\sqrt{12}}{2}$$

$$1 \pm \frac{2\sqrt{3}}{2} = 1 \pm \sqrt{3}$$

The only one of these which appears as an answer choice is $1 + \sqrt{3}$.

20. B: The graph of $y = x^2 + 2$ has a vertex at (0, 2) on the *y*-axis. The circle with a center at (0, 1) also lies on the *y*-axis. With a radius of 1, the circle touches the parabola at one point: the vertex of the parabola (0, 2).

Advanced Algebra and Functions

1. B: Start with the original equation: x- 2xy + 2y, then replace each instance of x with a 2, and each instance of y with a 3 to get:

$$2^2 - 2 \cdot 2 \cdot 3 + 2 \cdot 3^2 = 4 - 12 + 18 = 10$$

2. A: To expand a squared binomial, it's necessary to use the *First, Inner, Outer, Last Method*.

$$(2x - 4y)^2$$

$$(2x)(2x) + (2x)(-4y) + (-4y)(2x) + (-4y)(-4y)$$

$$4x^2 - 8xy - 8xy + 16y^2$$

$$4x^2 - 16xy + 16y^2$$

3. D: Factor the numerator into $x^2 - 6x + 9 = (x - 3)^2$, since:

$$-3 - 3 = -6$$

$$(-3)(-3) = 9$$

Factor the denominator into $x^2 - x - 6 = (x - 3)(x + 2)$, since:

$$-3 + 2 = -1,$$

$$(-3)(2) = -6$$

This means the rational function can be rewritten as:

$$\frac{x^2 - 6x + 9}{x^2 - x - 6} = \frac{(x - 3)^2}{(x - 3)(x + 2)}$$

Using the restriction of x > 3, do not worry about any of these terms being 0, and cancel an $x - 3$ from the numerator and the denominator, leaving $\frac{x-3}{x+2}$.

4. C: The area of the shaded region is the area of the square, minus the area of the circle. The area of the circle will be πr^2. The side of the square will be $2r$, so the area of the square will be $4r^2$. Therefore, the difference is:

$$4r^2 - \pi r^2 = (4 - \pi)r^2$$

5. B: Add 3 to both sides to get $4x = 8$. Then divide both sides by 4 to get $x = 2$.

6. B: There are two zeros for the given function. They are $x = 0, -2$. The zeros can be found several ways, but this particular equation can be factored into:

$$f(x) = x(x^2 + 4x + 4)$$

$$x(x + 2)(x + 2)$$

By setting each factor equal to zero and solving for x, there are two solutions. On a graph, these zeros can be seen where the line crosses the x-axis.

7. D: This problem involves a composition function, where one function is plugged into the other function. In this case, the $f(x)$ function is plugged into the $g(x)$ function for each x-value. The composition equation becomes:

$$g(f(x)) = 2^3 - 3(2^2) - 2(2) + 6$$

Simplifying the equation gives the answer:

$$g(f(x)) = 8 - 3(4) - 2(2) + 6$$

$$8 - 12 - 4 + 6 = -2$$

8. C: Graphing the function $y = \cos(x)$ shows that the curve starts at $(0, 1)$, has an amplitude of 2, and a period of 2π. This same curve can be constructed using the sine graph, by shifting the graph to the left $\frac{\pi}{2}$ units. This equation is in the form $y = \sin(x + \frac{\pi}{2})$.

9. D: Using SOHCAHTOA, tangent is $\frac{y}{x}$ for the special triangles. Since the value of $\tan(x)$ needs to be negative one, the angle for the tangent must be some form of 45 degrees or $\frac{\pi}{4}$. The value is negative in the second and fourth quadrant, so the answer is $\frac{3\pi}{4}$ and $\frac{7\pi}{4}$.

10. B: To simplify this inequality, subtract 3 from both sides to get:

$$-\frac{1}{2}x \geq -1$$

Then, multiply both sides by -2 (remembering this flips the direction of the inequality) to get $x \leq 2$.

11. D: There are two ways to approach this problem. Each value can be substituted into each equation. A can be eliminated, since:

$$4^2 + 16 = 32$$

Choice B can be eliminated, since:

$$4^2 + 4 \cdot 4 - 4 = 28$$

C can be eliminated, since

$$4^2 - 2 \cdot 4 - 2 = 6$$

But, plugging in either value into $x^2 - 16$, which gives:

$$(\pm 4)^2 - 16 = 16 - 16 = 0$$

12. A: The common denominator here will be $4x$. Rewrite these fractions as:

$$\frac{3}{x} + \frac{5u}{2x} - \frac{u}{4}$$

$$\frac{12}{4x} + \frac{10u}{4x} - \frac{ux}{4x}$$

$$\frac{12x + 10u - ux}{4x}$$

13. D: The expression is simplified by collecting like terms. Terms with the same variable and exponent are like terms, and their coefficients can be added.

14. D: Finding the zeros for a function by factoring is done by setting the equation equal to zero, then completely factoring. Since there was a common x for each term in the provided equation, that would be factored out first. Then the quadratic that was left could be factored into two binomials, which are $(x + 1)(x - 4)$. Setting each factor equal to zero and solving for x yields three zeros.

15. D: Dividing rational expressions follows the same rule as dividing fractions. The division is changed to multiplication by the reciprocal of the second fraction. This turns the expression into:

$$\frac{5x^3}{3x^2y} \times \frac{3y^9}{25}$$

Multiplying across and simplifying, the final expression is $\frac{xy^8}{5}$.

16. B: The table shows values that are increasing exponentially. The differences between the inputs are the same, while the differences in the outputs are changing by a factor of 2. The values in the table can be modeled by the equation $f(x) = 2^x$.

17. A: Using the trigonometric identity $\tan(\theta) = \frac{\sin(\theta)}{\cos(\theta)}$, the expression becomes $\frac{\sin\theta}{\cos\theta}\cos\theta$. The factors that are the same on the top and bottom cancel out, leaving the simplified expression $\sin\theta$.

18. C: The equation $x = 3$ is not a function because it does not pass the vertical line test. This test is made from the definition of a function, where each x-value must be mapped to one and only one y-value. This equation is a vertical line, so the x-value of 3 is mapped with an infinite number of y-values.

19. A: The function can be factored to identify the zeros. First, the term $3x$ is factored out to the front because each term contains $3x$. Then, the quadratic is factored into $(x + 3)(x - 2)$.

20. B: Plugging into the quadratic formula yields, for solutions:

$$x = \frac{1 \pm \sqrt{-1 + 4 \cdot 1(-3)}}{2} = \frac{1}{2} \pm \frac{\sqrt{11}}{2}$$

Therefore:

$$x - \frac{1}{2} = \pm\frac{\sqrt{11}}{2}$$

Now, if this is squared, then the $\pm$ cancels and left with:

$$\left(\frac{\sqrt{11}}{2}\right)^2 = \frac{11}{4}$$

Practice Test #2

Arithmetic

1. What is $\frac{660}{100}$ rounded to the nearest integer?
   a. 67
   b. 66
   c. 7
   d. 6

2. Which of the following is largest?
   a. -0.45
   b. -0.096
   c. -0.3
   d. -0.313

3. What is the value of *b* in this equation?

$$5b - 4 = 2b + 17$$

   a. 13
   b. 24
   c. 7
   d. 21

4. Katie works at a clothing company and sold 192 shirts over the weekend. One third of the shirts that were sold were patterned, and the rest were solid. Which mathematical expression would calculate the number of solid shirts Katie sold over the weekend?
   a. $192 \times \frac{1}{3}$

   b. $192 \div \frac{1}{3}$

   c. $192 \times \left(1 - \frac{1}{3}\right)$

   d. $192 \div 3$

5. Arrange the following numbers from least to greatest value:
   $0.85, \frac{4}{5}, \frac{2}{3}, \frac{91}{100}$

   a. $0.85, \frac{4}{5}, \frac{2}{3}, \frac{91}{100}$

   b. $\frac{4}{5}, 0.85, \frac{91}{100}, \frac{2}{3}$

   c. $\frac{2}{3}, \frac{4}{5}, 0.85, \frac{91}{100}$

   d. $0.85, \frac{91}{100}, \frac{4}{5}, \frac{2}{3}$

6. Divide and reduce $5/13 \div 25/169$.
    a. 13/5
    b. 65/25
    c. 25/65
    d. 5/13

7. $\frac{14}{15} + \frac{3}{5} - \frac{1}{30} =$
    a. $\frac{19}{15}$
    b. $\frac{43}{30}$
    c. $\frac{4}{3}$
    d. $\frac{3}{2}$

8. A piggy bank contains 12 dollars' worth of nickels. A nickel weighs 5 grams, and the empty piggy bank weighs 1050 grams. What is the total weight of the full piggy bank?
    a. 1,110 grams
    b. 1,200 grams
    c. 2,250 grams
    d. 2,200 grams

9. Last year, the New York City area received approximately $27\frac{3}{4}$ inches of snow. The Denver area received approximately 3 times as much snow as New York City. How much snow fell in Denver?
    a. 60 inches
    b. $27\frac{1}{4}$ inches
    c. $9\frac{1}{4}$ inches
    d. $83\frac{1}{4}$ inches

10. What is the solution to the following problem in decimal form?
$$\frac{3}{5} \times \frac{7}{10} \div \frac{1}{2}$$
    a. 0.042
    b. 84%
    c. 0.84
    d. 0.42

11. After a 20% discount, Frank purchased a new refrigerator for $850. How much did he save from the original price?
    a. $170
    b. $212.50
    c. $105.75
    d. $200

12. Jessica buys 10 cans of paint. Red paint costs $1 per can and blue paint costs $2 per can. In total, she spends $16. How many red cans did she buy?
    a. 2
    b. 3
    c. 4
    d. 5

13. If a car can travel 300 miles in 4 hours, how far can it go in an hour and a half?
    a. 100 miles
    b. 112.5 miles
    c. 135.5 miles
    d. 150 miles

14. At the store, Jan spends $90 on apples and oranges. Apples cost $1 each and oranges cost $2 each. If Jan buys the same number of apples as oranges, how many oranges did she buy?
    a. 20
    b. 25
    c. 30
    d. 35

15. A train traveling 50 miles per hour takes a trip lasting 3 hours. If a map has a scale of 1 inch per 10 miles, how many inches apart are the train's starting point and ending point on the map?
    a. 14
    b. 12
    c. 13
    d. 15

16. A traveler takes an hour to drive to a museum, spends 3 hours and 30 minutes there, and takes half an hour to drive home. What percentage of his or her time was spent driving?
    a. 15%
    b. 30%
    c. 40%
    d. 60%

17. Simplify the following fraction:

$$\frac{\frac{5}{7}}{\frac{9}{11}}$$

    a. $\frac{55}{63}$

    b. $\frac{7}{1000}$

    c. $\frac{13}{15}$

    d. $\frac{5}{11}$

18. Greg buys a $10 lunch with 5% sales tax. He leaves a $2 tip after his bill. How much money does he spend?
    a. $12.50
    b. $12
    c. $13
    d. $13.25

19. Marty wishes to save $150 over a 4-day period. How much must Marty save each day on average?
    a. $37.50
    b. $35
    c. $45.50
    d. $41

20. A couple buys a house for $150,000. They sell it for $165,000. By what percentage did the house's value increase?
    a. 10%
    b. 13%
    c. 15%
    d. 17%

Quantitative Reasoning, Algebra, and Statistics

1. What is the value of x in the following expression?
$$4x - 12 = -2x$$
    a. 2
    b. 3
    c. -6
    d. -2

2. What is the simplified quotient of the following expression?
$$\frac{5x^3}{3x^2y} \div \frac{25}{3y^9}$$
    a. $\frac{125x}{9y^{10}}$

    b. $\frac{x}{5y^8}$

    c. $\frac{5}{xy^8}$

    d. $\frac{xy^8}{5}$

3. The area of a given rectangle is 24 square centimeters. If the measure of each side is multiplied by 3, what is the area of the new figure?
    a. 48cm
    b. 72cm
    c. 216cm
    d. 13,824cm

4. In the figure below, what is the area of the shaded region?

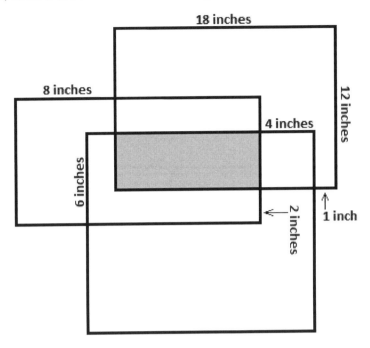

    a. 48 sq. inches
    b. 52 sq. inches
    c. 44 sq. inches
    d. 56 sq. inches

5. If the ordered pair $(-3, -4)$ is reflected over the x-axis, what's the new ordered pair?
    a. $(-3, -4)$
    b. $(3, -4)$
    c. $(3, 4)$
    d. $(-3, 4)$

6. What would the equation be for the following problem?
               3 times the sum of a number and 7 is greater than or equal to 32
    a. $3(7n) > 32$
    b. $3 \times n + 7 \geq 32$
    c. $3n + 21 > 32$
    d. $3(n + 7) \geq 32$

7. The table below shows tickets purchased during the week for entry to the local zoo. What is the mean of adult tickets sold for the week?

Day of the Week	Age	Tickets Sold
Monday	Adult	22
Monday	Child	30
Tuesday	Adult	16
Tuesday	Child	15
Wednesday	Adult	24
Wednesday	Child	23
Thursday	Adult	19
Thursday	Child	26
Friday	Adult	29
Friday	Child	38

a. 24.2
b. 21
c. 22
d. 26.4

8. Courtney leaves home and drives her truck 1,236 yards towards her destination. Her destination is 6,292 feet away from her home. How many more feet does she need to travel before she arrives?

a. 2,284

b. 3,708

c. 5,056

d. 2,584

9. A local candy store reports that of the 100 customers that bought suckers, 35 of them bought cherry. What is the probability of selecting 2 customers simultaneously at random that both purchased a cherry sucker?

a. $\frac{119}{990}$

b. $\frac{35}{100}$

c. $\frac{49}{400}$

d. $\frac{69}{99}$

10. Which inequality represents the number line below?

a. $4x + 5 < 8$
b. $-4x + 5 < 8$
c. $-4x + 5 > 8$
d. $4x - 5 > 8$

11. Which of the following formulas would correctly calculate the perimeter of a legal-sized piece of paper that is 14 inches long and $8\frac{1}{2}$ inches wide?

    a. $P = 14 + 8\frac{1}{2}$

    b. $P = 14 + 8\frac{1}{2} + 14 + 8\frac{1}{2}$

    c. $P = 14 \times 8\frac{1}{2}$

    d. $P = 14 \times \frac{17}{2}$

12. What is the slope of the line that passes through the points $(10, -4)$ and $(-5, 8)$?

    a. $-\frac{5}{4}$

    b. $-\frac{4}{15}$

    c. $-\frac{4}{5}$

    d. $-\frac{12}{5}$

13. Points L and M lie on line KN. The length of line KN is 30 units long, LN is 20 units long, and KM is 16 units long. How many units long is LM?

    a. 16
    b. 14
    c. 10
    d. 6

14. Angle y measures 48°. Angle x measures twice the value of angle y. What is the value of angle z?

    a. 96°

    b. 36°

    c. 132°

    d. 42°

15. What is the measurement of angle f in the following picture? Assume the lines are parallel.

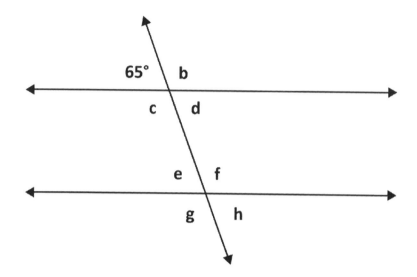

    a. 65 degrees
    b. 115 degrees
    c. 125 degrees
    d. 55 degrees

16. Given the triangle below, find the value of x if $y = 21$.

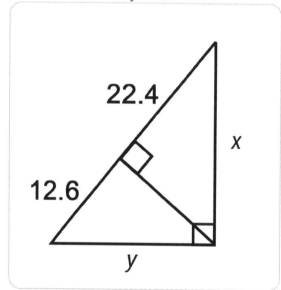

    a. 35
    b. 28
    c. 25
    d. 26

17. What is the value of x in the following triangle?

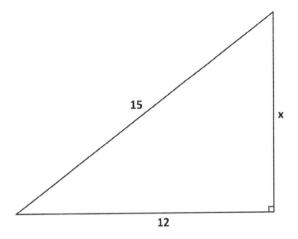

a. 19.2
b. 9
c. 3
d. 7.5

18. A data set is comprised of the following values: 30, 33, 33, 26, 27, 32, 33, 35, 29, 27. Which of the following has the greatest value?
    a. Mean
    b. Median
    c. Mode
    d. Range

19. A group of marathon runners were surveyed to see how many miles they ran in a week to prepare for the upcoming race. Six runners were surveyed, and the median of the responses was 14 miles. Which of the following is a possible list of responses?
    a. 11, 12, 15, 16, 11, 13
    b. 13, 16, 11, 12, 15, 16
    c. 16, 16, 14, 13, 11, 12
    d. 12, 15, 16, 13, 16, 15

20. 250 students were asked their favorite flavor of ice cream. The results are presented in the pie chart below. What is the probability of choosing a student who likes vanilla or strawberry ice cream to receive a free ice cream cone?

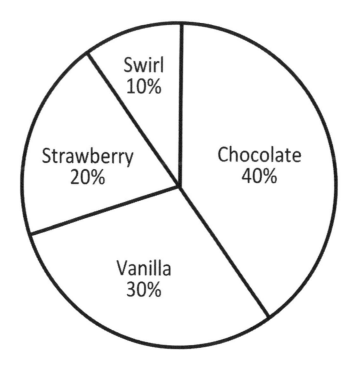

Favorite Ice Cream Flavor

a. $\frac{3}{10}$

b. $\frac{2}{10}$

c. $\frac{1}{2}$

d. $\frac{1}{20}$

Advanced Algebra and Functions

1. If $x = 2.6$ and $y = 5.3$, what is the value of $3.2xy - 4.1y$?
    a. -7.95
    b. 33.436
    c. 44.096
    d. 22.366

2. What is the solution to the radical equation $\sqrt[3]{2x + 11} + 9 = 12$?
    a. -8
    b. 8
    c. 0
    d. 12

3. What is the solution to the following system of equations?
$$\begin{cases} x^2 + y = 4 \\ 2x + y = 1 \end{cases}$$

   a. (–1, 3)
   b. (–1, 3), (3, –5)
   c. (3, –5)
   d. –1, 3

4. If $f(x) = x^2 - 3x + 17$, then what is $f(x + 1)$?
   a. $x^2 - 3x + 19$
   b. $x^2 - x + 15$
   c. $x^2 + 2x + 18$
   d. $x^2 - 3x + 14$

5. What is the product of the following expression?
$$(x + 2)(x^2 + 5x - 6)$$

   a. $8x^2 + 4x - 12$
   b. $x^2 + 6x - 4$
   c. $x^3 + 7x^2 + 4x - 12$
   d. $x^3 + 5x^2 - 4x - 12$

6. What are the roots of $x^2 + x - 2$?
   a. 1 and -2
   b. -1 and 2
   c. 2 and -2
   d. 9 and 13

7. What is the y-intercept of $y = x^{5/3} + (x - 3)(x + 1)$?
   a. 3.5
   b. 7.6
   c. -3
   d. -15.1

8. $(4x^2y^4)^{\frac{3}{2}}$ can be simplified to which of the following?
   a. $8x^3y^6$

   b. $4x^{\frac{5}{2}}y$

   c. $4xy$

   d. $32x^{\frac{7}{2}}y^{\frac{11}{2}}$

9. Which graph will be a line parallel to the graph of $y = 3x - 2$?
   a. $2y - 6x = 2$
   b. $y - 4x = 4$
   c. $3y = x - 2$
   d. $2x - 2y = 2$

10. A company invests $50,000 in a building where they can produce saws. If the cost of producing one saw is $40, then which function expresses the amount of money the company pays? The variable y is the money paid and x is the number of saws produced.
    a. $y = 50,000x + 40$
    b. $y + 40 = x - 50,000$
    c. $y = 40x - 50,000$
    d. $y = 40x + 50,000$

11. If $3x = 6y = -2z = 24$, then what does $4xy + z$ equal?
    a. 116
    b. 130
    c. 84
    d. 108

12. Which inequality represents the following number line?

    a. $-\frac{5}{2} \leq x < \frac{3}{2}$

    b. $-\frac{7}{2} \leq x < \frac{5}{2}$

    c. $-\frac{5}{2} < x \leq \frac{3}{2}$

    d. $\frac{5}{2} < x \leq -\frac{3}{2}$

13. What is the equation of a circle whose center is (0, 0) and whole radius is 5?
    a. $(x - 5)^2 + (y - 5)^2 = 25$
    b. $(x)^2 + (y)^2 = 5$
    c. $(x)^2 + (y)^2 = 25$
    d. $(x + 5)^2 + (y + 5)^2 = 25$

14. Which of the ordered pairs below is a solution to the following system of inequalities?

$$y > 2x - 3$$
$$y < -4x + 8$$

    a. (4, 5)
    b. (-3, -2)
    c. (3, -1)
    d. (5, 2)

15. Which equation best represents the scatterplot below?

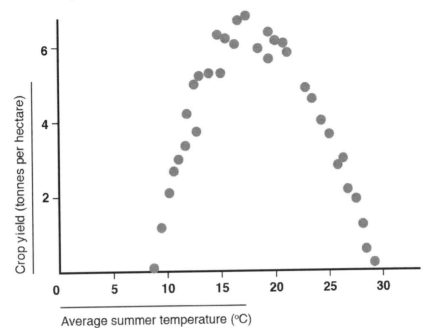

a. $y = 3x - 4$
b. $y = 2x^2 + 7x - 9$
c. $y = (3)(4^x)$
d. $y = -\frac{1}{14}x^2 + 2x - 8$

16. The triangle shown below is a right triangle. What's the value of x?

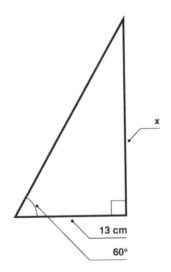

13 cm

60°

a. $x = 1.73$
b. $x = 0.57$
c. $x = 13$
d. $x = 22.52$

17. If $n = 2^2$, and $m = n^2$, then m^n equals?
  a. 2^{12}
  b. 2^{10}
  c. 2^{18}
  d. 2^{16}

18. The graph of the function $f(x) = |x + 3| - 4$ is graphed on the coordinate plane below.

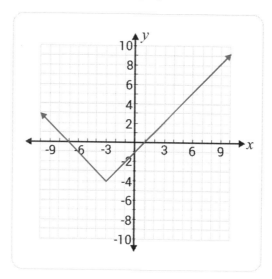

For which of the following functions does the graph of $g(x)$ intersect the function $f(x)$ exactly two times?
  a. $g(x) = x - 4$
  b. $g(x) = -x$
  c. $g(x) = x + 2$
  d. $g(x) = \frac{x}{2}$

19. If h is a multiple of 5, which of the following must also be multiples of 5?
  a. $5h - 3$
  b. $4h + 1$
  c. $10h + 20$
  d. $h^3 + 6$

20. Of the given sets of coordinates below, which ones lie on the line that is perpendicular to $y = 2x - 3$ and passes through the point $(0,5)$?
  a. $(-6,10)$
  b. $(-2,7)$
  c. $(2,6)$
  d. $(4,3)$

Answer Explanations #2

Arithmetic

1. C: Dividing by 100 means shifting the decimal point of the numerator to the left by 2. The result is 6.6 and rounds to 7.

2. B: Because these decimals are all negative, the number that is the largest will be the number whose absolute value is the smallest, as that will be the negative number with the least value. Thus, it will be the "least negative" (closest to zero). To figure out which number has the smallest absolute value, look at the first non-zero digits. The first non-zero digit in Choice *B* is in the hundredths place. The other three all have non-zero digits in the tenths place, so Choice *B* is closest to zero; thus, it is the largest of the four negative numbers.

3. C: To solve for the value of b, both sides of the equation need to be equalized.

Start by cancelling out the lower value of -4 by adding 4 to both sides:

$$5b - 4 = 2b + 17$$
$$5b - 4 + 4 = 2b + 17 + 4$$
$$5b = 2b + 21$$

The variable *b* is the same on each side, so subtract the lower 2b from each side:

$$5b = 2b + 21$$
$$5b - 2b = 2b + 21 - 2b$$
$$3b = 21$$

Then divide both sides by 3 to get the value of *b*:

$$3b = 21$$

$$\frac{3b}{3} = \frac{21}{3}$$

$$b = 7$$

4. C: $\frac{1}{3}$ of the shirts sold were patterned. Therefore, $1 - \frac{1}{3} = \frac{2}{3}$ of the shirts sold were solid. Anytime "of" a quantity appears in a word problem, multiplication should be used.

Therefore:

$$192 \times \frac{2}{3} = \frac{192 \times 2}{3} = \frac{384}{3} = 128 \text{ solid shirts were sold}$$

The entire expression is $192 \times \left(1 - \frac{1}{3}\right)$.

5. C: The first step is to depict each number using decimals. $\frac{91}{100} = 0.91$

Dividing the numerator by denominator of $\frac{4}{5}$ to convert it to a decimal yields 0.80, while $\frac{2}{3}$ becomes 0.66 recurring. Rearrange each expression in ascending order, as found in Choice *C*.

6. A: 13/5

Set up the division problem.

$$\frac{5}{13} \div \frac{25}{169}$$

Flip the second fraction and multiply.

$$\frac{5}{13} \times \frac{169}{25}$$

Simplify and reduce with cross multiplication.

$$\frac{1}{1} \times \frac{13}{5}$$

Multiply across the top and across the bottom to solve.

$$\frac{1 \times 13}{1 \times 5} = \frac{13}{5}$$

7. D: Start by taking a common denominator of 30.

$$\frac{14}{15} = \frac{28}{30}, \frac{3}{5} = \frac{18}{30}, \frac{1}{30} = \frac{1}{30}$$

Add and subtract the numerators for the next step.

$$\frac{28}{30} + \frac{18}{30} - \frac{1}{30} = \frac{28 + 18 - 1}{30} = \frac{45}{30} = \frac{3}{2}$$

In the last step the 15 is factored out from the numerator and denominator.

8. C: A dollar contains 20 nickels. Therefore, if there are 12 dollars' worth of nickels, there are $12 \times 20 = 240$ nickels. Each nickel weighs 5 grams. Therefore, the weight of the nickels is $240 \times 5 = 1,200$ grams. Adding in the weight of the empty piggy bank, the filled bank weighs 2,250 grams.

9. D: 3 must be multiplied times $27\frac{3}{4}$. In order to easily do this, the mixed number should be converted into an improper fraction.

$$27\frac{3}{4} = \frac{27 * 4 + 3}{4} = \frac{111}{4}$$

Therefore, Denver had approximately $\frac{3x111}{4} = \frac{333}{4}$ inches of snow. The improper fraction can be converted back into a mixed number through division.

$$\frac{333}{4} = 83\frac{1}{4} \text{ inches}$$

10. C: The first step in solving this problem is expressing the result in fraction form. Separate this problem first by solving the division operation of the last two fractions. When dividing one fraction by another, invert or flip the second fraction and then multiply the numerator and denominator.

$$\frac{7}{10} \times \frac{2}{1} = \frac{14}{10}$$

Next, multiply the first fraction with this value:

$$\frac{3}{5} \times \frac{14}{10} = \frac{42}{50}$$

In this instance, you can find the decimal form by converting the fraction into $\frac{x}{100}$, where x is the number from which the final decimal is found. Multiply both the numerator and denominator by 2 to get the fraction as an expression of $\frac{x}{100}$.

$$\frac{42}{50} \times \frac{2}{2} = \frac{84}{100}$$

In decimal form, this would be expressed as 0.84.

11. B: Since $850 is the price *after* a 20% discount, $850 represents 80% of the original price. To determine the original price, set up a proportion with the ratio of the sale price (850) to original price (unknown) equal to the ratio of sale percentage:

$$\frac{850}{x} = \frac{80}{100}$$

12. C: We are trying to find x, the number of red cans. The equation can be set up like this:

$$x + 2(10 - x) = 16$$

The left x is actually multiplied by $1, the price per red can. Since we know Jessica bought 10 total cans, $10 - x$ is the number blue cans that she bought. We multiply the number of blue cans by $2, the price per blue can.

That should all equal $16, the total amount of money that Jessica spent. Working that out gives us:

$$x + 20 - 2x = 16$$

$$20 - x = 16$$

$$x = 4$$

13. B: 300 miles in 4 hours is 300/4 = 75 miles per hour. In 1.5 hours, the car will go 1.5×75 miles, or 112.5 miles.

14. C: One apple/orange pair costs $3 total. Therefore, Jan bought 90/3 = 30 total pairs, and hence, she bought 30 oranges.

15. D: First, the train's journey in the real word is 3 x 50 = 150 miles. On the map, 1 inch corresponds to 10 miles, so there is 150/10 = 15 inches on the map.

16. B: The total trip time is 1 + 3.5 + 0.5 = 5 hours. The total time driving is 1 + 0.5 = 1.5 hours. So, the fraction of time spent driving is 1.5/5 or 3/10. To get the percentage, convert this to a fraction out of 100. The numerator and denominator are multiplied by 10, with a result of 30/100. The percentage is the numerator in a fraction out of 100, so 30%.

17. A: First simplify the larger fraction by separating it into two. When dividing one fraction by another, remember to *invert* the second fraction and multiply the two as follows:

$$\frac{5}{7} \times \frac{11}{9}$$

The resulting fraction $\frac{55}{63}$ cannot be simplified further, so this is the answer to the problem.

18. A: The tip is not taxed, so he pays 5% tax only on the $10. 5% of $10 is $0.05 \times 10 = \$0.50$. Add up $10 + $2 + $0.50 to get $12.50.

19. A: The first step is to divide up $150 into four equal parts. 150/4 is 37.5, so she needs to save an average of $37.50 per day.

20. A: The value went up by $165,000 − $150,000 = $15,000. Out of $150,000, this is $\frac{15,000}{150,000} = \frac{1}{10}$. Convert this to having a denominator of 100, the result is $\frac{10}{100}$ or 10%.

Quantitative Reasoning, Algebra, and Statistics

1. A: To solve for x the steps are as follows:

$$4x - 12$$

$$-2x, 6x - 12 = 0$$

$$6x = 12$$

$$x = 2$$

2. D: Dividing rational expressions follows the same rule as dividing fractions. The division is changed to multiplication by the reciprocal of the second fraction. This turns the expression into:

$$\frac{5x^3}{3x^2y} \times \frac{3y^9}{25}$$

Multiplying across and simplifying, the final expression is:

$$\frac{xy^8}{5}$$

3. C: 216cm. Because area is a two-dimensional measurement, the dimensions are multiplied by a scale that is squared to determine the scale of the corresponding areas. The dimensions of the rectangle are multiplied by a scale of 3. Therefore, the area is multiplied by a scale of 3^2 (which is equal to 9):

$$24 \ cm \times 9 = 216 \ cm$$

4. B: This can be determined by finding the length and width of the shaded region. The length can be found using the length of the top rectangle, which is 18 inches, then subtracting the extra length of 4 inches and 1 inch. This means the length of the shaded region is 13 inches. Next, the width can be determined using the 6 inch measurement and subtracting the 2 inch measurement. This means that the width is 4 inches. Thus, the area is:

$$13 \times 4 = 52 \; sq.\,in.$$

5. D: When an ordered pair is reflected over an axis, the sign of one of the coordinates must change. When it's reflected over the x-axis, the sign of the y-coordinate must change. The x-value remains the same. Therefore, the new ordered pair is $(-3, 4)$.

6. D: 3 times the sum of a number and 7 is greater than or equal to 32 can be translated into equation form utilizing mathematical operators and numbers.

7. C: To find the mean, or average, of a set of values, add the values together and then divide by the total number of values. Each day of the week has an adult ticket amount sold that must be added together. The equation is as follows:

$$\frac{22 + 16 + 24 + 19 + 29}{5} = 22$$

8. D: First, convert the distance that Courtney already drove to feet. Because there are three feet per yard, her distance traveled thus far in yards must be multiplied by 3:

$$1,236 \times 3 = 3,708 \text{ feet}$$

If the total distance to travel is 6,292 feet, there is $6292 - 3708 = 2,584$ feet left to travel.

9. A: The probability of choosing two customers simultaneously is the same as choosing one and then choosing a second without putting the first back into the pool of customers. This means that the probability of choosing a customer who bought cherry is $\frac{35}{100}$. Then without placing them back in the pool, it would be $\frac{34}{99}$.

So, the probability of choosing 2 customers simultaneously that both bought cherry would be:

$$\frac{35}{100} \times \frac{34}{99}$$

$$\frac{1,190}{9,900}$$

$$\frac{119}{990}$$

10. B: The number line shows:

$$x > -\frac{3}{4}$$

Each inequality must be solved for x to determine if it matches the number line. Choice A of $4x + 5 < 8$ results in $x < -\frac{3}{4}$, which is incorrect. Choice C of $-4x + 5 > 8$ yields $x < -\frac{3}{4}$, which is also incorrect. Choice D of $4x - 5 > 8$ results in $x > \frac{13}{4}$, which is not correct. Choice B, $-4x + 5 < 8$ is the only choice that results in the correct answer of:

$$x > -\frac{3}{4}$$

11. B: The perimeter of a rectangle is the sum of all four sides. Therefore, the answer is:

$$P = 14 + 8\frac{1}{2} + 14 + 8\frac{1}{2}$$

$$14 + 14 + 8 + \frac{1}{2} + 8 + \frac{1}{2} = 45 \text{ square inches.}$$

12. C: The equation used to find the slope of a line when given two points is as follows:

$$slope = \frac{y_2 - y_1}{x_2 - x_1}$$

Substituting the points into the equation yields:

$$\frac{8 - (-4)}{-5 - 10}$$

$$\frac{12}{-15}$$

$$-\frac{4}{5}$$

13. D: The length of LM can be found by a series of calculations:

$$KL + LM = 16 \qquad\qquad KL = 16 - LM$$

$$LM + MN = 20 \qquad\qquad MN = 20 - LM$$

$$KN = KL + MN + LM = 30$$

$$16 - LM + 20 - LM + LM = 30$$

$$36 - 30 = LM$$

$$6 = LM$$

14. B: The three angles lie on a straight line; therefore, the sum of all the angles must equal $180°$. The values for angle x and angle y should be added together and subtracted from $180°$ to find the value for angle z as follows:

$$180 - \left(48° + 2(48°)\right) = 36°$$

15. B: Because the 65-degree angle and angle *b* sum to 180 degrees, the measurement of angle *b* is 115 degrees. Because of corresponding angles, angle *b* is equal to angle *f*. Therefore, angle *f* measures 115 degrees.

16. B: This triangle can be labeled as a right triangle because it has a right-angle measure in the corner. The Pythagorean Theorem can be used here to find the missing side lengths. The Pythagorean Theorem states that $a^2 + b^2 = c^2$, where a and b are side lengths and c is the hypotenuse. The hypotenuse, c, is equal to 35, and 1 side, a, is equal to 21. Plugging these values into the equation forms:

$$21^2 + b^2 = 35^2$$

Squaring both given numbers and subtracting them yields the equation:

$$b^2 = 784$$

Taking the square root of 784 gives a value of 28 for b. In the equation, b is the same as the missing side length x.

17. B: This problem can be solved using the Pythagorean Theorem. The triangle has a hypotenuse of 15 and one leg of 12. These values can be substituted into the Pythagorean formula to yield:

$$12^2 + b^2 = 15^2$$

$$144 + b^2 = 225$$

$$81 = b^2$$

$$b = 9$$

In this problem, b is represented by x so $x = 9$ is the correct answer.

18. C: Each value can be calculated so that they can be compared to find which one is the greatest. The mean is equal to:

$$\frac{26 + 27 + 27 + 29 + 30 + 32 + 33 + 33 + 33 + 35}{10} = 30.5$$

The median is equal to:

$$\frac{30 + 32}{2} = 31$$

The mode is equal to 33 because that number occurs 3 times in the data set. The range is equal to:

$$35 - 26 = 9$$

Therefore, the mode is the greatest value of the answer choices.

19. B: When the number of data points provided is an even number, then the average of the two middle points is the median. Each set of responses provided should be ordered from least to greatest, and then the middle two values should be averaged together to see which set provides a median of 14. Choice *B,* when ordered, is 11, 12, 13, 15, 16, and 16. The middle two values averaged together is $\frac{13+15}{2} = 14$ miles, which is the correct answer.

20. C: These two events are mutually exclusive because the students only picked one flavor of ice cream as the favorite so a student can't have chosen two flavors. Therefore, the probability of choosing a student who likes each flavor of interest (vanilla and strawberry) should be added together. 30% of students chose vanilla, and 20% of the students chose strawberry. Expressed as percentages the probabilities are $\frac{3}{10}$ and $\frac{2}{10}$ which can be added together to find:

$$\frac{3}{10} + \frac{2}{10} = \frac{5}{10} = \frac{1}{2}$$

Advanced Algebra and Functions

1. D: The values for x and y should be plugged into the equation to find the correct answer.

$$3.2(2.6)(5.3) - 4.1(5.3) = 22.366$$

2. B: First, subtract 9 from both sides to isolate the radical. Then, cube each side of the equation to obtain:

$$2x + 11 = 27$$

Subtract 11 from both sides, and then divide by 2. The result is $x = 8$. Plug 8 back into the original equation to obtain the true statement to check the answer:

$$\sqrt[3]{16 + 11} + 9$$

$$\sqrt[2]{27} + 9$$

$$3 + 9 = 12$$

3. B: The system can be solved using substitution. Solve the second equation for *y*, resulting in:

$$y = 1 - 2x$$

Plugging this into the first equation results in the quadratic equation:

$$x^2 - 2x + 1 = 4$$

In standard form, this equation is equivalent to $x^2 - 2x - 3 = 0$ and in factored form is:

$$(x - 3)(x + 1) = 0$$

Its solutions are $x = 3$ and $x = -1$. Plugging these values into the second equation results in $y = -5$ and $y = 3$, respectively. Therefore, the solutions are the ordered pairs $(-1, 3)$ and $(3, -5)$.

4. B: The function presented is being evaluated for $x + 1$; therefore, $x + 1$ must be substituted into the original function as follows:

$$f(x + 1) = (x + 1)^2 - 3(x + 1) + 17$$

The squared portion of the function becomes $x^2 + 2x + 1$, and distributing the -3 results in:

$$f(x + 1) = x^2 + 2x + 1 - 3x - 3 + 17$$

Combining like terms results in:

$$x^2 - x + 15$$

5. C: Finding the product means distributing one polynomial to the other so that each term in the first is multiplied by each term in the second. Then, like terms can be collected. Multiplying the factors yields the expression:

$$x^3 + 5x^2 - 6x + 2x^2 + 10x - 12$$

Collecting like terms means adding the x^2 terms and adding the x terms. The final answer after simplifying the expression is:

$$x^3 + 7x^2 + 4x - 12$$

6. A: Finding the roots means finding the values of x when y is zero. The quadratic formula could be used, but in this case, it is possible to factor by hand, since the numbers -1 and 2 add to 1 and multiply to -2. So, factor $x^2 + x - 2 = (x - 1)(x + 2) = 0$, then set each factor equal to zero. Solving for each value gives the values $x = 1$ and $x = -2$.

7. C: To find the y-intercept, substitute zero for x, which gives us:

$$y = 0^{\frac{5}{3}} + (0 - 3)(0 + 1)$$

$$0 + (-3)(1) = -3$$

8. A: Simplify this to:

$$(4x^2y^4)^{\frac{3}{2}} = 4^{\frac{3}{2}}(x^2)^{\frac{3}{2}}(y^4)^{\frac{3}{2}}$$

$$4^{\frac{3}{2}} = (\sqrt{4})^3 = 2^3 = 8$$

For the other, recall that the exponents must be multiplied; this yields:

$$8x^{2 \cdot \frac{3}{2}}y^{4 \cdot \frac{3}{2}} = 8x^3y^6$$

9. A: Parallel lines have the same slope. The slope of C can be seen to be 1/3 by dividing both sides by 3. The others are in standard form $Ax + By = C$, for which the slope is given by $\frac{-A}{B}$. The slope of A is 3, the slope of B is 4. The slope of D is 1.

191

10. D: For manufacturing costs, there is a linear relationship between the cost to the company and the number produced, with a y-intercept given by the base cost of acquiring the means of production, and a slope given by the cost to produce one unit. In this case, that base cost is $50,000, while the cost per unit is $40. So:

$$y = 40x + 50,000$$

11. A: First solve for x, y, and z. So:

$$3x = 24$$

$$x = 8$$

$$6y = 24$$

$$y = 4$$

$$-2z = 24$$

$$z = -12$$

This means the equation would be $4(8)(4) + (-12)$, which equals 116.

12. A: The solid dot is located between -2 and -3, and the open dot is located between 1 and 2. Therefore, x is between -2.5 and 1.5, which can be converted to $-\frac{5}{2}$ and $\frac{3}{2}$. The solid dot indicates greater than or equal to, and the open dot indicates less than so the inequality is:

$$-\frac{5}{2} \leq x < \frac{3}{2}$$

13. C: Nothing is added to x and y since the center is 0 and 5^2 is 25. Choice A is not the correct answer because you do not subtract the radius from x and y. Choice B is not the correct answer because you must square the radius on the right side of the equation. Choice D is not the correct answer because you do not add the radius to x and y in the equation.

14. B: For an ordered pair to be a solution to a system of inequalities, it must make a true statement for BOTH inequalities when substituting its values for x and y. Substituting (-3,-2) into the inequalities produces $(-2) > 2(-3) - 3$, which is $-2 > -9$, and $(-2) < -4(-3) + 8$, or $-2 < 20$. Both are true statements.

15. D: The shape of the scatterplot is a parabola (U-shaped). This eliminates Choices A (a linear equation that produces a straight line) and C (an exponential equation that produces a smooth curve upward or downward). The value of a for a quadratic function in standard form ($y = ax^2 + bx + c$) indicates whether the parabola opens up (U-shaped) or opens down (upside-down U). A negative value for a produces a parabola that opens down; therefore, Choice B can also be eliminated.

16. D: SOHCAHTOA is used to find the missing side length. Because the angle and adjacent side are known, $\tan 60 = \frac{x}{13}$. Making sure to evaluate tangent with an argument in degrees, this equation gives:

$$x = 13 \tan 60 = 13 \times \sqrt{3} = 22.52$$

17. D: If $n = 2^2, n = 4$, and $m = 4^2 = 16$. This means that $m^n = 16^4$. This is the same as 2^{16}.

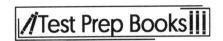

18. D: Each of the choices given is a linear function in the form of $y = mx + b$, where m is the slope, and b is the y-intercept. For the first function $g(x) = x - 4$, the slope is 1, and the y-intercept is -4. Placing this line on the graph would show that it does not intersect the absolute value function above. It crosses the y-axis at -4 with a slope equal to the right side of the absolute value graph, so they run parallel to one another. The second function will intersect the graph at 1 point because the y-intercept is 0. The line will run parallel to the left side of the absolute value function, with a slope of -1. For the third function, the slope is 1, and the y-intercept is 2. It will cross the absolute value graph at 1 point because it runs parallel to the right side of the graph and runs through the y-axis at 2. The fourth function has a slope of ½ and a y-intercept of 0. This line will intersect the absolute value graph at exactly 2 points. One point is at (2, 1), and the other point will be between x-values of -4 and -5. This line is shown on the graph below.

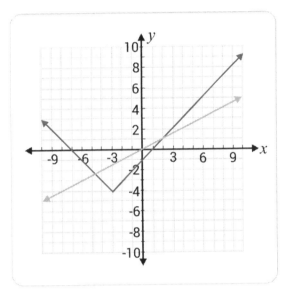

19. C: The value of $5h - 3$ cannot be a multiple of 5 because of the minus 3 at the end of the expression. Multiplying h by 5 yields a multiple of 5, but subtracting 3 means the final answer will no longer be a multiple of 5. For $10h + 20$, both the number multiplied by h and the number added to h are divisible by 5. The result will always be divisible by 5. For the final expression $h^3 + 6$, the value of h cubed yields an answer divisible by 5, but the addition of 6 means the answer is no longer divisible by 5.

20. D: The first step is to find the equation of the line that is perpendicular to $y = 2x - 3$ and passes through the point $(0, 5)$. The slope of a perpendicular line is found by the negative reciprocal of 2, which is $-\frac{1}{2}$. The y-intercept is the value of y when $x = 0$, so the y-intercept is 5. The new equation is $y = -\frac{1}{2}x + 5$. In order to find which points lie on the new line, the values of x and y can be substituted into the equation to determine if they form a true statement. For the point in A, the equation $10 = -\frac{1}{2}(-6) + 5$ makes the statement $10 = 8$. This is not a true statement, so the point $(-6, 10)$ does not lie on the line. For B, the equation $7 = -\frac{1}{2}(-2) + 5$ makes the statement $7 = 6$, which is not a true statement. Therefore, Choice B is not a point that lies on the line. For Choice C, the equation $6 = -\frac{1}{2}(2) + 5$ is the same as $6 = 4$, so it is similarly false. For D, the equation $3 = -\frac{1}{2}(4) + 5$ makes a true statement, so the point $(4, 3)$ lies on the line.

Dear ACCUPLACER Math Test Taker,

We would like to start by thanking you for purchasing this study guide for your ACCUPLACER exam. We hope that we exceeded your expectations.

Our goal in creating this study guide was to cover all of the topics that you will see on the test. We also strove to make our practice questions as similar as possible to what you will encounter on test day. With that being said, if you found something that you feel was not up to your standards, please send us an email and let us know.

We would also like to let you know about other books in our catalog that may interest you.

SAT

This can be found on Amazon: amazon.com/dp/1628456108

ACT

amazon.com/dp/1628458844

We have study guides in a wide variety of fields. If the one you are looking for isn't listed above, then try searching for it on Amazon or send us an email.

Thanks Again and Happy Testing!
Product Development Team
info@studyguideteam.com

FREE Test Taking Tips DVD Offer

To help us better serve you, we have developed a Test Taking Tips DVD that we would like to give you for FREE. **This DVD covers world-class test taking tips that you can use to be even more successful when you are taking your test.**

All that we ask is that you email us your feedback about your study guide. Please let us know what you thought about it – whether that is good, bad or indifferent.

To get your **FREE Test Taking Tips DVD**, email freedvd@studyguideteam.com with "FREE DVD" in the subject line and the following information in the body of the email:

a. The title of your study guide.

b. Your product rating on a scale of 1-5, with 5 being the highest rating.

c. Your feedback about the study guide. What did you think of it?

d. Your full name and shipping address to send your free DVD.

If you have any questions or concerns, please don't hesitate to contact us at freedvd@studyguideteam.com.

Thanks again!

Made in United States
Orlando, FL
14 April 2024

45720265R50113